♦ GENES AND GENDER VI ♦

On Peace, War, and Gender: A Challenge to Genetic Explanations

The Genes and Gender series takes its shape and subject—the critique of genetic determinism—from the diverse work of the activists and scientists who make up the Genes and Gender Collective and contribute to the published proceedings of its conferences. Organized in 1977, the collective advocates responsible scientific and social research and challenges theories holding that genetic processes are primary determinants of human behavior and explain differences.

◆ GENES AND GENDER VI ◆

On Peace, War, and Gender: A Challenge to Genetic Explanations

EDITED BY ANNE E. HUNTER

Associate Editors, Catherine M. Flamenbaum and Suzanne R. Sunday
Series Editors, Betty Rosoff and Ethel Tobach

THE FEMINIST PRESS
at The City University of New York
New York

Published 1991 by The Feminist Press at The City University of New York,
311 East 94th Street, New York, NY 10128
Distributed by The Talman Company, 150 Fifth Avenue, New York, NY
10011

94 93 92 91 6 5 4 3 2 1

Library of Congress Cataloging-in-Publication Data

Genes and gender VI: on peace, war, and gender: a challenge to genetic explanations / edited by Anne E. Hunter; associated editors.
p. cm.—(Genes and gender; 6)
Based on proceedings of the Sixth Genes and Gender Conference held at Stern College, Yeshiva University in 1986.
Includes bibliographical references.
ISBN 1-55861-037-5: $35.00. — ISBN 1-55861-025-1 (pbk.): $12.95
1. Aggressiveness (Psychology)—Genetic aspects. 2. War—Psychological aspects. 3. Peace—Psychological aspects. 4. Sex differences (Psychology) 5. Women—Psychological aspects. 6. Sociobiology. I. Hunter, Anne E. II. Flamenbaum, Catherine M. III. Sunday, Suzanne R. IV. Genes and Gender Conference (6th: 1986: Stern College, Yeshiva University. V. Title: Genes and gender 6. VI. Series.
BF341.G39 no. 6
[BF575.A3]
155.2′34 s—dc20
[303.6′6]

90-14107
CIP

Cover design: Gilda Hannah
Text design: Paula Martinac

Printed in the United States of America on acid-free paper by McNaughton & Gunn, Inc.

The Feminist Press gratefully acknowledges the gift of Letty and Bert Pogrebin toward publication of this book. Publication has also been made possible, in part, by public funds from the New York State Council on the Arts.

This volume is dedicated to Eleanor Burke Leacock,
founding member of the Genes and Gender Collective:
scientist, teacher, scholar,
and inspiring leader of all who worked with her

Contents

♦ PART II ♦
Assigning Gender to Peace and War: Social Processes

♦ PART III ♦
Who Fights for Peace and Who Makes War?

Foreword

This book, the sixth in the Genes and Gender series, is the result of a Genes and Gender Conference on peace and war held at Stern College, Yeshiva University, in 1986. It was the last Genes and Gender Conference in which Eleanor Burke Leacock, a founder of the Genes and Gender Collective, participated before her death, and the book is dedicated to her.

Two events stimulated the organization of this conference. One was an article by Kay Camp and an answering letter by Mamie Jackson, both appearing in *Peace and Freedom*, the publication of the Women's International League for Peace and Freedom. The second was a presentation by the Burlington Women's Collective at a meeting of the Association for Women in Psychology. The ideas expressed by these women made it clear that there were many points of view about the causes of peace and war. It seemed appropriate, therefore, for Genes and Gender to take up the issues of peace and war and expose the pseudoscientific justification for a genetic determinist view of the causes of militarism.

In preparing the papers for publication, we soon realized that there were other topics that should be considered within the discussion. We invited papers by Daisy Garth on women's peace work in Nicaragua, Sally Kitch on the connection between war and gender roles, and Doris and Erwin Marquit on social and economic reasons for women's pacifism. As Eleanor Leacock could not complete the writing of her paper, we asked Geraldine Casey, one of her students and a member of the

Genes and Gender Collective, to write on the topic that Leacock had presented at the conference.

We want to thank the editors, Anne Hunter, Catherine Flamenbaum, and Suzanne Sunday, for all the work they did in organizing the conference and carrying the volume through to completion. We thank all those who contributed to the volume either at the conference or later. We also wish to thank the Dean of Stern College, Karen Bacon, who graciously offered the facilities of the college to the Biology Club, which cosponsored the conference with the Genes and Gender Collective.

Ethel Tobach and Betty Rosoff

Preface

In the tradition of the Genes and Gender series, this volume continues to challenge genetic explanations of human behavior and social organization and to expose the political uses of this pseudoscientific theory. The theory of genetic determinism—most systematically embraced by E. O. Wilson, biologist and founder of the field of sociobiology—holds two major assumptions: that all aspects of human behavior and culture are coded in the genes, and that the particular genes forming the roots of human culture have been evolutionarily selected because they lead to optimal reproductive fitness among those who have them. Under the auspices of science, therefore, the theory sanctions hierarchical social arrangements (social classes, patriarchy, white supremacy, and imperialism) by arguing that they are both unavoidable and evolutionarily adaptive.

The goal of the current book is to critically evaluate the application of sociobiology to the cultural manifestations of war and aggression. Specifically, the book challenges the position that warfare is inevitable by evaluating the two major assumptions underlying it: that men are by nature territorial and competitive, and that they are genetically programmed to dominate women, who are by nature passive and nurturing. By its deterministic thrust, sociobiology asserts an extremely pessimistic view of human nature, implying that organized movements for national and world peace are naive and futile.

Alternatively, the Genes and Gender perspective on human nature is one of optimism and choice—that is, that human cultural manifestations such as warfare and peace are rooted in human choice and possibility rather than genetic fixity. Our complex brains allow us to choose between war and peace, rather than forcing men to aggress and women to nurture. To support this optimistic theory of human nature, the book provides evidence for the following points: (1) warfare is not universal among humans; (2) women and men have equal potential for making war and peace; and (3) peacemaking and war making cannot be explained by simplistic, reductionistic, and linear formulas such as those represented by sociobiology. Instead, peace and war arise out of a complicated interplay among genes, biology, society, and culture, which means that humans and other animals are constantly transforming and being transformed by their internal and external environments. Thus, they are both the results and causes of their environments.

What is important about this dialectical view of human nature is that it asserts world peace as a human possibility, a potential equally available to men and women. The key to unlocking our peacemaking potential, however, lies in an understanding of the broader social context of human lives. Growing cooperation among countries to halt the further destruction of our planet is testament to the human capacity to act in ways that are mutually rewarding rather than competitive. The recent abatement of the threat of nuclear war between the United States and the Soviet Union is further testament. And, women's activism in all parts of the world against social injustice challenges genetic theories concerning the innate passivity of females. Unfortunately, the simultaneous growth of nationalism in Europe and the continued growth of racism and chauvinism internationally are threatening to undermine these life-promoting, cooperative activities. These contradictory social behaviors make the work of the Genes and Gender Collective as significant as ever, for, if we cannot envision a world built on shared power and cooperation, we will never take the steps to achieve it.

The critical question becomes: Under what social conditions will our biological and uniquely human capacity for *choosing* peace over war outweigh our tendency for choosing war over peace? The essays in this book, written from the perspectives of many disciplines, address this question.

Anne Hunter

♦ GERALDINE J. CASEY AND ETHEL TOBACH ♦

Eleanor Burke Leacock: A Tribute

Eleanor Burke was born on July 2, 1922, and she died on April 2, 1987 while on sabbatical from the Anthropology Department of City College, the City University of New York. She was studying the educational and social lives of the people of western Samoa as they were relevant to the differing views held by Derek Freeman and Margaret Mead, a debate then engaging the interests of the scientific world. She believed that genetic determination underlay Freeman's views and that the interests of the people of West Samoa, as well as those of many peoples who had been "studied" by anthropologists, needed to be central to that debate.

Leacock received her bachelor's degree in 1944 and her master's degree in 1946 from Columbia University. She received her doctorate from the same institution in 1952. During those years she married Richard Leacock and had four children, all of whom accompanied her on her precedent-setting field research for her doctorate. She was a lecturer and teacher at the Polytechnic Institute of Brooklyn, Queens College, New York University, and at many institutions in Europe. She was appointed to the faculty of the Anthropology Department at City College in 1972 and directed the research of many students who were fulfilling the requirements for the master's and doctorate degrees.

Eleanor Leacock was an internationally renowned anthropologist, a committed teacher, a theoretical pathbreaker, and a consistent activist in the struggles against racism, sexism, and any form of genetic determinism in science. Her doctoral thesis reconstructed the history of

Indians (the Montagnais-Naskapi of Labrador), challenging the idea that private property was universal and demonstrating the existence of precapitalist social equality. Her introduction to Friedrich Engels's *The Origin of the Family, Private Property, and the State* is recognized as the germinal work reintroducing Marxism into anthropology.

Leacock conducted other fieldwork on racism in the New York City public schools, housing desegregation in New Jersey, education and social policy in Zambia, and ethnohistory and adolescence in Samoa. She is perhaps best known for her antiracist work attacking the "culture of poverty" theory. Her research in public education demonstrated that teacher expectations, often unconsciously racist and sexist, affected student performance more than any cultural attributes students brought into the classroom. In the 1970s Eleanor Leacock pioneered research and developed theory on women's inequality and the struggle for empowerment. Here again she rejected genetic arguments to explain historically constituted social inequality.

She was a member of many national and international anthropological and scientific associations, winning recognition for her scholarship and support for other, younger scholars in the field of anthropology. In 1983 Leacock became the first woman to win the New York Academy of Sciences Award in the Behavioral Sciences.

She was also a prolific scholar, and the editor, coeditor, or author of thirteen books and monographs, more than eighty articles or chapters in books, and numerous book reviews, comments, and invited lectures. Many of her works have been translated into foreign languages.

Professor Leacock was a role model for her students and younger colleagues. She consistently fought to make academia accessible to disenfranchised people, and she understood that their participation would strengthen the rigor and depth of scientific theory and research. Most important, she understood that gaining access to existing structures was not enough—that a transformational process was required to remake our ways of knowing, of conducting research, and of teaching in order to serve the communities we study and build collective action. Eleanor Leacock was a partisan scholar, an engaged activist, and a committed friend to all those people whose lives she touched.

♦ GENES AND GENDER VI ♦

On Peace, War, and Gender: A Challenge to Genetic Explanations

♦ GERALDINE J. CASEY ♦

Eleanor Leacock, Marvin Harris, and the Struggle over Warfare in Anthropology

In 1986, Eleanor Leacock addressed the following questions at the Genes and Gender conference on war and peace: What contribution can anthropology make to our understanding of warfare? And how do the distortions embedded in various theories of warfare and human violence serve to perpetuate a gendered concept of war and peace (1986a)? This paper will examine two opposed theoretical approaches to these issues as expressed in the work of Eleanor Leacock and Marvin Harris.

Harris, a well-known anthropologist, has written extensively on warfare in relation to what he calls the "Male Supremacist Complex" (Harris 1972, 1974, 1977, 1984, 1989; and Divale and Harris 1976, 1978a, 1978b). Harris argues that warfare is a universal feature of and a functional adaptation for human society; it is a mechanism for regulating population growth in relation to the resources available in the environment.

Several people read and commented on an earlier version of this paper. I want to thank Dr. Leith Mullings and Dr. Jane Schneider from the Anthropology Department at the CUNY Graduate Center for their encouragement to pursue the project and for their helpful editorial comments. Dr. Ethel Tobach, Eleanor Leacock's close friend and colleague for many years, was a source of constant intellectual insight, challenge, and support throughout the course of writing and editing the paper. In addition, I am extremely grateful for the questions, criticisms, and the ongoing political and historical discussion provided by Rachel Kranz, Bill Askins, and Paul Mishler.

At the conference Eleanor Leacock presented a critique of Marvin Harris's ideas. She refuted the scientific content and logic of his arguments and explored the subtle, albeit powerful, ideological pressures that compel many social scientists, including Marvin Harris, to bring their research and writing into line with the specific cultural values and political assumptions that dominate contemporary United States society. Leacock believed that warfare emerges at specific points in human history and is the result of complex social pressures, not the result of genetic structures (1979, 1981a, 1982c, 1986a). In her presentation Leacock presented a wide range of ethnographic and historical data that contradict the idea that warfare is innate, universal, and genetically determined. Leacock provided anthropologists with an alternative framework from which to study the origins and structures of war. Her emphasis on gender issues highlighted the roles of women in periods of war and in the struggle for peace.

Leacock argued that it is important to consider the political-economic role of social scientists in society in order to understand how academic theories of warfare and women's roles are developed. She applied a Marxist theoretical framework to show how anthropological studies of warfare have served to reinforce dominant social values (1982b). Leacock described how political pressures and career aspirations confront social scientists and "encourage" them, in direct and subtle ways, to bring their research into alignment with the ideological tenets of genetic determinism and sociobiology:

> Like the social sciences generally, anthropology has always been torn between science and ideology: between the goal of achieving a scientific understanding of society in order to improve it, and the requirement of our present social system for the ideological supports that help maintain it. (1986a)

Leacock consistently placed her research topics in historical perspective in order to trace the relationship between academic scholarship and social practice. She was particularly concerned about the way that academic theories are applied by governments to rationalize specific policies and social programs. (In the context of our discussion on gendered and genetic-determinant theories of warfare, the policy applications seem very tangible and dangerously concrete.) Leacock emphasized the way that institutionalized class, racial, and gender hierarchies function to maintain their vested interests and to squelch alternative, critical approaches. Using a sociology-of-knowledge approach, she traced the relationship between supposedly pure research and the construction of "common sense" knowledge and social meaning:

> The influence of our particular social-economic structure, and the myriad ways in which the carrot and stick are used to discourage any challenge to the status quo, cause anthropological data to be manipulated to fit rationales for injustice and inequality, for discrimination, and for the prevalence of war and violence. (1986a)

One major objective of this paper is to examine the contributions of Eleanor Leacock's scholarship to the continuing struggle against theories of genetic determinism and for the cause of peace and women's full equality.[1]

MARVIN HARRIS AND THE ANTHROPOLOGY OF WAR

Marvin Harris has produced seventeen books, including several best-sellers and a popular introductory college textbook. His work has been translated into fifteen languages. The publishers at Harper and Row describe Marvin Harris as "one of the best known of contemporary anthropologists [who] has devoted his career to asking big questions about the human condition and answering them authoritatively in clear, down-to-earth prose" (Harris 1989, 548). War is certainly one of the "big questions," and Marvin Harris has written a great deal about it (Harris 1972, 1974, 1977, 1981, 1989; and Divale and Harris 1976, 1978a, 1978b).

The Early Years, 1968: Antiwar Efforts, Racial Justice, and Women's Rights

Harris was one of the editors of the first comprehensive study of warfare ever produced in anthropology. In their introductory essay the editors describe the process that led to the book's publication: the volume emerged from an "impromptu midnight caucus of anthropologists opposed to the United States Viet Nam policy" at the 1966 Annual Meeting of the American Anthropological Association (AAA) (Fried, Harris, and Murphy 1968, xvi). A number of anthropologists were "indignant over the lack of attention" being paid to the Vietnam War. Overnight, activists were able to draft and distribute a petition that, by the next morning, had 350 names attached to it. The petitioners demanded that an "unprecedented" plenary symposium on the anthropology of warfare be scheduled for the annual meeting in 1967. Harris was a member of the ad hoc committee that took responsibility for organizing the sessions and editing the resulting volume for publication.

In addition to making plans for the symposium on warfare, the 1967 AAA annual meeting passed the following resolution:

> Reaffirming our 1961 Resolution, we condemn the use of napalm, chemical defoliants, harmful gases, bombing, the torture and killing of prisoners of war and political prisoners, and the intentional or deliberate policies of genocide or forced transportation of populations for the purpose of terminating their cultural and/or genetic heritages by anyone anywhere.
>
> These methods of warfare deeply offend human nature. We ask that all governments put an end to their use at once and proceed as rapidly as possible to a peaceful settlement of the war in Viet Nam. (Fried, Harris, and Murphy 1968, xvi)

The symposium on warfare took place on November 30, 1967. By December the papers from that session were compiled and edited, and the book was published in 1968. The schedule indicates the degree of commitment mobilized by antiwar anthropologists and their sense of urgency. The entire process underscores Leacock's observation that political pressures in the historical moment (in this case, pressure from the political Left) shape the production of scientific scholarship.

A wide range of data was presented. Paleontology and human evolution, archaeology, physical anthropology, and social/cultural anthropology were represented. The volume included case study reports and more general papers that attempted to develop theories on the causes and functions of war. A record of the sharp debates that took place at the session was also included. The editors reported that a consensus was reached on certain key issues that have a special relevance for this essay. Regarding the definition of *warfare*, the editors stressed the need to maintain a clear distinction between types of armed conflict that take place "among remote peoples in the Orinoco jungles of Amazonia and New Guinea" and warfare in complex, urban societies:

> Yanomamo "warfare" compares with youthful gang fighting or brawls between individuals which are never considered "war" in our culture. While in complex society, one state attacks another with 100,000 troops, aircraft, warships, yet cannot say for sure that a war does nor does not "exist" because certain secular rituals, i.e., a congressional declaration of war, have not yet been held. (Fried, Harris, and Murphy 1968, xvii–xviii)

The book raised questions that inform our discussion of war today, more than twenty years later: What is the definition of war? How can we distinguish conflict, aggression, and violence from warfare? Does war serve some function? If so, does that quality make it somehow "adaptive" for human society? How significant are the issue of scale and the differences between episodes of organized violence in small-scale, pre-industrial societies and the modern warfare of military superpowers?

Harris was part of an ascending generation of anthropologists who set out to challenge many of the discipline's basic tenets. His volume on the history of anthropological theory had just come out (1968). In it, Harris criticized the dominant conceptual framework in anthropology. He rejected what he called "cultural idealism" and "mentalism" in mainstream anthropology, a theoretical approach that claimed that patterns of human behavior are generated in the mind, completely divorced from the impact of environment, technological resources, and social organization. Harris pointed out that this approach fractured the integrated nature of human experience into isolated domains and tried to elevate the cultural realm above the political and economic spheres of existence. He dismissed the idea that the entire range of human social variation could be explained solely on the basis of different cultural beliefs and values. In a later work Harris projected an alternative theory of "cultural materialism," which postulated that environmental constraints, population pressure, and different levels of technological sophistication constitute the prime movers in human history (1979). He argued that differences in these forces account for the wide range of human social and cultural variation.

In 1968 Harris was perceived as a radical reformer. He argued that as a discipline, anthropology developed in the late nineteenth century as a response to Marxism. Anthropology projected an idealist view of human history in order to refute the impact of Marx's vision of class struggle and competing interests (1968, 249). Eleanor Leacock always acknowledged this aspect of Harris's early writings (Leacock 1972a, 61–66) and her article, on "Marxism and Anthropology," begins with a commentary on Harris's role in bringing Marxist concepts into anthropological discourse (1982b, 242). Leacock continued to recognize Harris's historical contribution: "He is a liberal who has criticized racism in anthropology, who has declared himself in strong support of women's equality, and who has not been afraid to credit the importance of Marx as a social scientist" (1986a).

Harris endorsed Marx's emphasis on the material forces of production, but he rejected the Marxist concept of dialectics as an "idealist" formulation and dismissed the Marxist emphasis on negation and social contraction in the development of human history. In accordance with Marx, Harris stressed the relationship between ecology and human economy, or the "forces of production," but Harris disagreed with the idea that the "*social* relations of production" also had a generative impact on the course of human history. For him relations between people, including organized political struggle, were always secondary—if not epiphenomenal—in relation to the power exerted by the demography and the environment. According to Harris, technological innovations are the only mechanisms that allow human societies to transcend the material constraints of increasing population measured against a limited resource base.

The more Harris developed his theory of cultural materialism, the more rigid and simplistic his writing became. He reduced the complex processes of social development to single units of technological innovation such as the introduction of the plow or the development of hoe agriculture. According to Harris, cultural materialism began to be dominant "when our ancestors crossed the threshold of cultural takeoff" (1989, 64). It was the moment when "natural selection had brought the body, brain and behavior of our ancestors to cultural takeoff, [and] culture itself began to evolve." Supposedly, the event occurred when "modern sapiens entered Europe and achieved cultural takeoff 45,000–35,000 years ago" (Harris 1989, 89, 126).

In Harris's view "cultural selection is human nature's servant." It replaces natural genetic selection to achieve the same ends: "Once cultural takeoff has occurred and cultural selection is operating at full force, differential reproductive success ceases to be the means by which variations . . . are propagated" (1989, 127).

This line of thinking is riddled with faulty assumptions. First of all, Harris accepts the sociobiologists' proposal that there is a set of innate, universal human needs. He also accepts their definition of a fixed "human nature," which they characterize as aggressive, competitive, and selfish. Harris merely emends the genetic determinists' argument by adding that, at a certain point in human cultural development, technological innovations replace natural selection in order to maintain the species' fitness and adaptation to the environment: "Innate aggressive potentials must surely be part of human nature . . . but cultural selection wields the power that activates or inactivates these raw potentials" (1989, 296).

The complexity and significance of the actual technological innovations are also reduced. Every innovation is assumed to serve a specific purpose in the maintenance and smooth functioning of the social system. Each is simply part of the "materialist strategies for optimization" that facilitate the search for "short-term and long-term trade-offs" in the struggles engendered by increasing population and a diminishing resource base (Harris 1987c, 74, 77).

Harris's theory of human culture cannot account for those features that are only partially integrated into the social whole, dysfunctional, or differentially functional (serving the interests of some sectors of society and not others); nor does it deal with cultural traits that are archaic holdovers from an earlier period yet serve no current purpose. The flexibility, fluidity, and openness of human cultures are reduced to technological mechanisms with precise functions.

Leacock criticized Harris's cultural materialism and his emphasis on environmental conditions and technology to the exclusion of other social forces (1972a, 61–66). She explained that he was "splitting the dialectic" between the social relations of production and material forces of production. His approach, she argued, led Harris to a "frankly Malthusian stance," creating an "assertive economic determinism" (1986a). Thus, Harris became a mechanical materialist and a technological reductionist, which, for Leacock, created serious problems in his analysis of warfare (1982, 266–67).

Marvin Harris on the Warpath

Initially, Harris's views were in accordance with those expressed in the 1968 AAA volume on warfare. He agreed that modern warfare was qualitatively different from the organized violence that took place in non–state-level societies, as among the Tiwi in Australia. In these contexts warfare was personalistic, ritually circumscribed in time and space, and came to a halt as soon as someone was injured. Harris also argued against biological determinism and the idea that warfare was an innate human drive.

Then, in 1976, Harris formed a partnership with William Tulio Divale and reversed his earlier position (Divale and Harris 1976, 1978a, 1978b). He began to consider war a natural and universal feature of human society. During the period of his collaboration with Divale, Harris narrowed his analysis of warfare. He came to focus exclusively on how it relates to caloric intake and the quest for more proteins—a question of "balancing population against resources" (1989, 296). Divale and Harris tried to prove their point by distorting and manipulating demographic statistics from a wide range of ethnographic and archaeological reports (see Diener, Moore, and Mutaw 1980; Howe 1978). Harris's most recent writings continue to argue that "preindustrial peoples mostly go to war to moderate or cushion the impact of unpredictable food shortages" (1989, 297). He uses many numbers and technical language in his writing to appear more "scientific." But all the quantitative analysis does not substantiate his claims, as Harris admits that there is no direct correlation between warfare and protein levels: "Yanomami villages with low levels of protein intake (thirty-six grams) seem to engage in warfare just as frequently as those that have high protein intake (seventy-five grams) per adult" (1989, 311).

Conjuring up a new phrase, Divale and Harris "invented the 'Male Supremacist Complex' " (Norton 1976), proposing that institutionalized female subordination was also functional and causally linked to the practice of warfare. They presented the following scenario: in order to keep population levels adaptive within the given ecological and technological framework, societies practice warfare and female infanticide to reduce male and female population levels, respectively. Warfare requires exaggerated male sexual aggression, and the entire syndrome supposedly functions as the Male Supremacist Complex (Divale and Harris 1976):

> Male tasks during times of warfare are simply more critical to the survival of everyone than is female work . . . Males were selected for the role of warriors because the anatomical and physiological sex-linked differences that favored the selection of men as hunters of animals favored the selection of men as hunters of people. In combat with hand-held weapons dependent on muscle-power, the slight 10 or 15 percent edge that men hold over women in track and field performances becomes a life-and-death issue, while the restrictions imposed on women by pregnancy loom as an even greater handicap in war than in hunting. (1989, 284)

Using a functionalist logic and a very simple cause-and-effect model, Harris has been quick to embrace "sex-linked differences" and genetic determinism as explanations of women's social inequality. Although Harris never provides adequate evidence to document that female infanticide is, indeed, an institutionalized practice, he insists that it is ubiquitous and a "functional adaptation:"

> The fecundity of the human female is so great that even if raiding reduces the [population] density of a territory by half, all it takes is twenty-five years of unrestrained breeding for the population to reach its former level. . . . Warfare places a premium on maximizing the number of future warriors, which leads to the preferential treatment of male infants and higher rates of direct and indirect female infanticide. (1989, 297)

Marvin Harris has come to view warfare as a fundamental part of *all* human society, including the more egalitarian band- and village-level societies, which predated the rise of the state and continue to exist in parts of the Third World. This position constitutes a reversal of his earlier stance in the 1968 volume on the anthropology of war. Divale and Harris dismiss the prevalent anthropological interpretation that links intensified, institutionalized warfare with the rise of the state. This approach is supported by Marxist and feminist research, which links the institutionalization of women's social inequality and warfare with the rise of class-based state societies.

Disregarding the volume of carefully researched ethnographic evidence, Harris affirms "the importance of warfare in shaping gender hierarchies among band-and-village peoples" (1989, 285). According to Harris, "many band-level societies do engage in moderate amounts of warfare and have correspondingly more pronounced forms of male sexism" (1989, 288). "Whenever conditions favored the intensification of warfare among bands and villages, the political and domestic subordination of women also became more intense" (1989, 284).

Harris insists that village- and band-level warfare predates colonial contact and continues into the present time, whenever and wherever the restraining or 'civilizing" influence of European colonial domination is unable to penetrate (1989, 324). To argue that intergroup conflict in the more environmentally and politically marginal sections of today's Third World nations has nothing to do with the ruptures and strains caused by the long period of European colonization and violent domination of these areas is to disregard history. Harris's second assertion, that this type of warfare is only curbed by the civilizing presence of those same colonizers, is a pro-imperialist statement. His viewpoint stands in direct opposition to the vast amount of ethnographic evidence amassed during the twentieth century. Harris's position perpetuates the racist image of tribal people as "savages" engaged in a desperate pursuit of sex, violence, and proteins.

At the Genes and Gender conference in 1986, Leacock elaborated on the twists and turns in Harris's position on warfare, claiming it had become "biological determinism in cultural clothing."

Harris had shifted his views considerably since the days of 1968 and the early AAA research on warfare, and evaluating his revised position, Leacock observed that

> Harris would certainly not see himself as a biological determinist. However, he has formulated a theory of warfare, violence, and male dominance that leans heavily on Malthus and on the influence of biology, and that in certain respects parallels sociobiology. In my view Harris's position can best be characterized as "biological determinism in cultural clothing." (1986a)

Harris changed his position on the role of warfare in human society at the same time that he changed his attitude toward women. Again, the new attitude developed subsequent to his involvement with Divale and their reification of warfare as an adaptive, technological strategy. At that point Harris began to adopt an attitude of gratuitous disrespect toward women, both in his writing style and in the content of his arguments. This stance is reflected in his descriptions of women in the ethnographic literature and his discussion of feminist colleagues in anthropology. It permeates Harris's popular book *Cannibals and Kings* (1977), especially the chapters on "The Origins of War" (1977, 31–43) and "The Origin of Male Supremacy and of the Oedipus Complex" (1977, 55–66). I offer two examples from his description of warfare among the Tiwi people of Australia:

> Not infrequently the person hit was some innocent noncombatant or one of the screaming old women who weaved through the fighting men, yelling obscenities at everybody.
>
> As soon as somebody was wounded, even a seemingly irrelevant crone, fighting stopped. (quoted in Harris 1975, 35)

The pictures of "screaming old women" and a "seemingly irrelevant crone" constitute powerful negative images. While it trivializes Tiwi women, the passage adds nothing to our understanding of Tiwi social behavior. The phrases that ridicule Tiwi women bear no connection to the argument at hand. Quite the contrary, we are informed that these women play an active role in the proceedings, even as noncombatants; should one of the women be wounded, all fighting comes to a halt. According to Harris's own report, Tiwi women seem to enjoy a position of importance denied to participants in modern warfare.

In the end Harris comes full circle. He joins the establishment group he regarded once with disdain. Subsequent to his partnership with Divale, Harris seems to embrace the same rhetoric used by Desmond Morris and Robert Ardrey: that human aggression is universal and inevitable: that warfare is adaptive; and that women are subordinate.

Harris publicly recanted his earlier position in the second edition of his popular college textbook, *People, Culture and Nature*: "There seems to be little basis for the view to which I myself subscribed in the previous edition of this book, that war is essentially a post-Neolithic phenomenon" (1975, 263). His disrespect for women is extended to include his colleagues in anthropology and feminist researchers as well:

> The practice of warfare is responsible for a widespread complex of male supremacist institutions among band and village societies. The existence of this complex is a source of embarrassment and confusion to advocates of women's rights. . . . Unfortunately, feminists have tried to counter the view that male supremacy is natural by denying that it existed among the majority of band and village peoples. . . . Feminists have recently insisted that [it] is an illusion created by the sexist minds of the male observers. . . . I find it impossible to believe that the overwhelming

> statistical regularities indicative of virtually universal structural biases against women are nothing more than motes in the eyes of male field workers. (1975, 57–58)

Harris likes to portray himself as the "objective" scientist who is not afraid to "shoot from the hip" and "tell it like it is." If the truth about warfare and women's subordination embarrasses or confuses "advocates of women's rights," that's just too bad. It is not at all clear, however, that Harris's "overwhelming statistical regularities" really exist. In reality, the promise of statistical proof may be nothing more than "motes in the eyes" of Divale and Harris.

This issue lies at the heart of the anthropological critique of Divale and Harris. Researchers who collected the primary source data have publicly refuted the ways that Divale and Harris used their materials (see, for example, Hirschfeld, Howe, and Levin 1978; Howe 1978; Lancaster and Lancaster 1978; and Norton 1978). The critics argue that Divale and Harris distorted and misrepresented the evidence on warfare to fit their theory. Howe provides explicit documentation to prove that Divale and Harris fabricated data for nonexisting communities (1978). In other circumstances they excluded data from the same research sample when it failed to support their hypothesis.

What factors can help to explain this situation? Why would Marvin Harris, a committed "man of science," violate the accepted standards for scholarship? The drive for prestige and authority in one's chosen field can exert a powerful pressure on academics, especially when it is mixed with the desire to be provocative, controversial, and confrontational. Moreover, William Tulio Divale had been an FBI informer during the 1960s, a fact whose implications Eleanor Leacock urged people to recognize (1986a). His role was to infiltrate and disrupt leftist and progressive political movements in California. It was his job to distort information and fabricate data to entrap people and substantiate baseless claims made by the FBI. The Divale-Harris writings on warfare and women's subordination makes use of the same "research" tactics.

Collaborating with a "Collaborateur"

Leacock said it was necessary to consider exactly *who* was William Tulio Divale (1986a). She explained that Divale had worked as a paid FBI informer for four years, from 1965 to 1969. In line with her commitment to trace out the connections between scholarship and sociopolitical action, Leacock argued that Divale's past practice should be made known because it was immoral and violated the established guidelines for anthropological behavior. She also believed there was a connection between his work as an FBI informer and his approach to scientific research, an approach that involved distortion and fabrication of the "evidence" to prove an ideological commitment.

Divale's autobiographical book, *I Lived Inside the Campus Revolution*, provides a firsthand report of this period (1970). As an undercover agent and provocateur, Divale was assigned to infiltrate and disrupt the Students for a Democratic Society (SDS) and the Communist party in California. His job was to create conflict and disrupt the daily activities of the anti-war movement, civil rights organizations, and the student movement. In addition to reporting everything he heard and saw to the FBI, Divale fabricated stories, forged documents, spread lies about activists, and set them up for arrest. One of his claims to fame is that he was the person who turned Angela Davis in to the FBI (1970, 163–87). In all his efforts to wreak havoc and disunity, Divale showed no respect for human rights: "In many ways I was a product of the Federal Bureau of Investigation. The Bureau had 'recruited' me when I was a political virgin. At the time I had no politics of my own and little of anyone else's" (1970, ix).

Ideological beliefs about the need to fight communism or defend America played little part in his decision to become a spy. Divale admits in print that he received at least $14,596, in 1960s dollars, for his efforts (1970, 165), and brags about his activities as a paid informer and agent provacateur:

> I had filed nearly eight hundred undercover FBI reports. . . . My undercover file for the Bureau ran to thousands of pages, documenting day by day, meeting to meeting. . . . I had named names—the names of perhaps four thousand activists. . . . I was a leader, not merely a follower. . . . In time, perhaps correctly, the FBI would come to list me among its 'most valuable' student informers. (1970, viii–ix)

In the pursuit of more fame and notoriety subsequent to his break with the FBI, Divale went public with his stories in a torrid, confessional memoir. Divale listed again, this time for public consumption, all the sexual affairs, family fights, and personal crises of the individuals he had previously spied on.

Because the issues of female subordination and male sexual aggression feature so prominently in Divale's subsequent writing on warfare, it is significant to examine his attitude toward women, as reflected in the memoir of his life as an FBI informer. Divale describes his sexual affairs at great length: "I'd slept with the Movement, taken to my bed its most fervent philosophy and some of its most passionate women" (1970, 143); "that summer of 1966 I put a lot more miles than women under me" (1970, 76).

Divale's descriptions of women, and human sexuality in general, are disrespectful and objectifying. Writing about a period of great social upheaval in the United States, the time of the so-called sexual revolution and the rise of a new women's movement in the 1960s, when human sexual relations, always complex, were being reexamined and challenged in new ways, Divale trivializes women's struggle for equality:

> Some Commie coeds aren't sure they are—or even want to be girls. . . . She looks female, dresses female, but she's all up-tight with inner aggressions and with the outward kind, too. If you watch her carefully, . . . you can all but see her feminine and masculine chromosomes flailing away at one another. (1970, 75–76)

His discussion of sexual relations is focused exclusively on women. Following the pattern of most male sociobiologists, he pays little attention to the roles of men in these relationships and to the impact of male supremacist ideology on the construction and distortion of gender roles. Sounding like a sociobiologist, Divale reduces the whole process to a war between "feminine and masculine chromosomes."

After his sordid career as an informer and pulp author, Divale went on to graduate school, earning a Ph.D. in anthropology. His status and reputation never seemed to suffer as a result of his ignoble actions. On the contrary, Divale seems to have received all the perquisites and benefits academia has to offer. He is currently chair of the Anthropology Department at York College, part of the City University of New York (CUNY) in Queens, New York.

In this light it is interesting to note that Eleanor Leacock was denied full access to academic resources and employment opportunities, the result of her radical politics and Marxist/feminist intellectual commitments. Gailey's 1988 biographical article on Leacock documents that after completing her Ph.D., it took Leacock eleven years to secure a full-time position teaching anthropology because, "in addition to being female, Leacock was married, she was a mother, and was a politically engaged radical" (Gailey 1988, 218).

Leacock never became disillusioned or cynical in response to these conditions, and she always maintained an active involvement in both anthropology and the struggle for social justice. Nevertheless, the contrast between her career path and that of Divale underscores Leacock's point about the influence of bourgeois ideology within the halls of academe. Academic institutions clearly reward those scholars whose work supports the dominant interests of the state to maintain the status quo, regardless of how sordid and dishonest their practices may be.

Eleanor Leacock insisted that any discussion of Divale's research on warfare and women's status must consider his history of personal and political duplicity as an FBI agent (1986a). Anthropologists have long considered spying for the government to be a violation of professional ethics. Franz Boas, one of the founders of anthropology, wrote a scathing letter condemning any social scientist who spied for the government, even during a period of declared war (Leacock 1982b, 244). Collusion of this sort has been criticized soundly by anthropologists ever since. During the Vietnam war, activist anthropologists condemned their colleagues who lent their research skills to aid the U.S. military in Thailand (Wolf and Jorgensén 1970).[2]

Yet, William Tulio Divale took money for spying. He fabricated and distorted information to serve his own interests and please his backers. His 1970 autobiography describes the opportunism that motivated him. It is highly unlikely that these tendencies disappeared once he decided to pursue an academic career in anthropology. This could explain why Divale might find it easy to fabricate ethnographic data and distort the evidence on village- and band-level societies in an effort to "prove" that they engaged in warfare and institutionalized female infanticide.

There may be an interesting thematic connection between Divale's undercover FBI work and the focus of his academic work on warfare and female subordination. The FBI assigned him to infiltrate and disrupt the 1960s student movements against racism, sexism, and the Vietnam war, and his academic theories postulate the universality of inequality, women's subordination, and warfare in human society.

ELEANOR LEACOCK'S CRITIQUE OF HARRIS

When Marvin Harris joined the sociobiologists to proclaim that human aggression is an innate, heritable feature and that warfare is an inherent part of "human nature," Eleanor Leacock rejected his claims, based on the anthropological evidence. She considered his actions to be in line with a more general pattern: "the continual emergence of theories that see warfare and violence as expressing innate human aggression, such as new versions of Robert Ardrey's 'territorial imperative' " (1986a).

Reactionary ethologists such as Robert Ardrey achieved notoriety and popular success in the 1960s for books that argued that all primates are aggressive, competitive, and organized for hierarchical fashion (1961); since humans are primates, we are only "naked apes" (Morris 1961), with all the same innate drives. Because this approach justifies contemporary forms of social inequality and imperialist aggression as "natural" phenomena, new versions of the old theme keep emerging.

As we have seen, Harris's particular variation merely replaced genetic structures with technological innovations and environmental constraints, while retaining the basic theme of a fixed human nature intact.

Academics who search for commonalities between different human cultures and between people and nonhuman primates conflate different units of analysis and different levels of abstraction into a homogenized whole. This approach leads to an overaggregation of disparate elements. It also produces facile scientific "theories" that offer simplistic, reductionist answers to complex social processes. Eleanor Leacock rejected this enterprise, relying upon the critical theories developed by Marxist and feminist scholars to develop a more historically accurate social science and a more rigorous approach to cross-cultural studies.

Harris and the sociobiologists believe that human beings are programmed to a very narrow range of behaviors. In contrast, Leacock explored the wide range of social forms in time (through archaeology and human evolution) and space (through cross-cultural comparison) that demonstrate the openness and variety of human cultures.

Like Harris, Leacock studied the "big questions." She was committed to the search for general patterns in human society—an integral part of the Marxist and feminist project in anthropology. As an authentic generalist, however, Leacock paid rigorous attention to historical specificity in her study. She never treated information gathered about a troop of macaque monkeys as though it was interchangeable with data on urban communities of people in the United States. Nor would she treat all sources of data equally, ignoring the historical and ideological interests they might reflect. In all of her anthropological studies, beginning with her doctoral research, for example, Leacock made extensive use of ethnohistorical materials. She scrutinized the contents of these documents and mined their richness, always paying careful attention to the ideological subtexts contained in the letters and reports of early colonial adventurers and administrators.

Divale, Harris, and Derek Freeman (another sociobiologist whose work will be discussed below) treat all their sources of data equally, making no distinction between the sermons and letters of missionaries, the ideological reports of colonial governors, and the more careful observations of recent ethnographers. It is all "just data" to them. Leacock's research traced the development of particular relationships in context, noting the points of convergence and the junctures of divergence in social development, rather than lumping together unlike phenomena in a superficial search for universal features and common trends.

Review of the Anthropological Data

Leacock's discussion of warfare at the Genes and Gender Conference began with a review of the literature:

> A wealth of anthropological data on the history, the cultures, the languages, and the physical characteristics of hundreds of different tribes and nations, past and present, offers a laboratory for the study of social processes and their relations with biological ones. (Leacock 1986a)

Leacock synthesized data from all four subdisciplines of anthropology to build a multifaceted picture of the role of violence in human society that contradicts the sociobiologists' theories (Fedigan 1986; Goodale 1971; Lee 1979; Rohrlich-Leavitt 1975; Slocum 1975). She integrated evidence from paleontology and archaeology, primate studies, research on extant groups of foragers, and historical studies. The information and insight gleaned from this information creates a powerful mosaic, one that disassembles the sociobiologists' claim that violence is an innate, genetically determined human characteristic (Leacock 1977d, 1978a, 1979 and 1981a)

Primate Studies. The image of nonhuman primates made famous by ethnologists like Ardrey (1961) and Morris (1967) describes them as aggressive, male-dominant, and living in hierarchically stratified groups. Since humans are members of the primate order, these attributes are also ascribed to them. It is a distorted picture of primate social life, however, and it reinforces simplistic notions about human aggression and violence.

Through the 1960s primatologists were overwhelmingly male, and their research reflected a definite male bias. They concentrated on the behavior of primate males, attending to issues of group dynamics only in situations of population stress or danger. They stressed relations of male dominance and authority and only dealt with female primates in the context of sexual relations.

Women researchers entered the field of primate studies in the late 1960s and 1970s. They began to ask different questions and their studies challenged those earlier assumptions. Researchers such as Jane Lancaster (1973, 1975), Lila Leibowitz (1979), and Ruby Rohrlich-Leavitt (1975) expanded the focus of primate study to include females, infants, and juvenile primates. They examined the range of daily activities, including communication, foodgathering, play, sharing, and cooperation. Their findings emphasized the variation among individuals in groups. The study of social relations moved beyond the mother-infant dyad and male-female sexual relations to include an examination of the relations between females, between juveniles and infants, and between males and juveniles. Based on this new research, Leacock concluded that "the human ancestor that emerges from primate data viewed in their entirety is an intensely sociable, communicative, curious, and playful creature, capable of widely ranging behaviors" (Leacock 1981a, 189).

The Archaeological Record. Archaeological and paleontological studies underwent a similar transformation during this time period. New questions were raised about the degree of social complexity and women's status in early human society (see, for example, Fedigan 1986; Leacock 1981a; Tanner and Zihlman 1976; and Zihlman and Tanner 1978). Leacock stressed the important corrective to notions about "Man the Hunter" that resulted from the examination of the realm of female activity, especially food gathering, food sharing, and socialization, in the study of hominid evolution.

The notion that early human males enjoyed a significant and absolute physical "advantage" over females was used by male researchers, including Harris, to explain how males achieved greater social status due to their importance in the hunting of large animals. Physiological differences in height, bones, and muscle power between men and women were assumed to lead directly to a total domination of men over women. Leacock refuted this unsubstantiated idea, citing the work of Erik Trinkaus on Neanderthal populations (1978). There was little sexual dimorphism among Neanderthals. Both women and men were heavily muscled, which, according to Leacock, implied that women would be able to participate equally with men in the task of digging pit traps for large mammals (Leacock 1981a, 190).

The archaeological information on early human societies that specifically addresses the issue of war and institutionalized violence strongly supports the position that warfare only emerged in the context of state building and long-distance trade, not among earlier gatherers and hunters living in bands and villages. Studying the cultures of Crete and Sumer, Ruby Rohrlich-Leavitt observed that "the development of trade and the specialization of labor led to politically organized urban societies and systematic warfare for access to trade routes and important resources" (1977). June Nash's research on Aztec society corroborates this trajectory (1978). She argues that Aztec warfare only developed in relation to state building and social-class stratification, and she traces the connection between male warfare, the state, and women's subordination—"the interrelationships between male specialization in warfare, predatory conquest, state bureaucracy . . . and the differential access to its benefits between men and women" (Nash 1978).

All of this more recent research challenges the established image of early humans as brutish, warlike, and controlled by innate biological "drives," or instincts. In her conference presentation Leacock concluded that

> the archeological and paleontological evidence of human social and physical evolution indicates cooperative groups with, on the whole, good relations with their neighbors; intermarrying, feasting, and helping out in cases of starvation. (1986a)

Contemporary Gatherers-Hunters. For more than 90 percent of our time on this planet, humans have lived as foragers, or gatherers/hunters (Leacock 1982a; Leacock and Lee 1982; Lee 1979; and Slocum 1975). This is the earliest and most enduring form of human social organization. If aggression and violence were innate human traits, we should expect to find evidence of them in the archaeological and historical record and contemporary ethnographic research on these communities.

Based on a thorough review of the research on these societies, Leacock asserted that "gathering-hunting societies, the type of society that has characterised by far the longest part of human existence, is the most peaceful type of society; not without problems, but they have *social* strategies for dealing with them" (Leacock 1986a). If violence is inherently human, Leacock asks, why is the most durable form of human society so peaceful?

Leacock's framework was grounded in her fieldwork research among the Montagnais-Naskapi of Labrador (Leacock 1955, 1958, 1969b, 1976b, 1980c; Leacock and Goodman 1976; Leacock and Rogers 1981). She was also involved in more generalized research with other scholars who studied band-level societies, such as Richard Lee (Leacock and Lee 1982), Nancy Lurie (Leacock and Lurie 1988), Edward Rogers, (Leacock and Rogers 1981), and Jacqueline Goodman (Leacock and Goodman 1976). The studies demonstrate the depth and breadth of Leacock's expertise in the study of foragers or band-level societies. On the basis of all this research, Leacock concluded:

> Sociobiology and theories of male aggression and female subservience flounder on the simple fact that men and women are more equal in gatherer-hunter societies (which is closer to some "basic" human nature) than in some more complex societies. (Leacock 1986a)

Historical Study versus Distortions of the Ethnographic Record. One of Leacock's major quarrels with sociobiologists and their supporters in anthropology, such as Harris, was with their ahistorical approach to the study of human society. Perhaps this follows from the sociobiologists' logic that, if warfare and aggression are natural, universal, and constant, what need is there to be historically specific? Leacock soundly rejected this approach. She cited Goodale's 1971 on Tiwi women. Goodale draws very different conclusions about women's roles in Tiwi society than those developed by Harris. Reflecting also on her own historical research on the Montagnais-Naskapi society in the seventeenth century, Leacock concluded that, "even with the stepping-up of warfare, following the land pressure caused by [European] colonization, the warfare was very different from what we know" (Leacock 1986a).

The tendency in ethnographic studies of non–state-level societies until the 1960s was to ignore the available historical records of initial European encounters with native peoples. Prominent anthropologists in Britain and the United States dismissed attempts at historical reconstruction as unscientific and speculative (Harris 1968; Leacock, 1972a, 1982b; Sider 1974). Historical study was assumed to be dangerously aligned with evolutionism and Marxism by way of Lewis Henry Morgan (1963) and Frederick Engels (see Leacock 1971b). Both American particularists and British structural functionalists preferred synchronic, empirical study based on what they would actually see on the ground. According to Leacock (1982b, 243–44), such studies were static, ahistorical, and overly empiricist. Because the research questions and theoretical models used in these projects lacked any grounding in the history of the societies under study, the interpretations and conclusions tended to reflect the basic assumptions of Victorian European culture and society, rather than presenting an accurate assessment of social life in the foraging societies of Africa, Asia, and Latin America.

Early ethnographers projected a fractured reflection of their own social values and anxieties onto the societies they studied. The aggression, competition, individualism, and male supremacy that investigators "discovered" in their fieldwork settings constituted more accurately a portrayal of the dominant ideologies and colonial policies of governments in Europe and the United States toward these people (for more about anthropology and colonialism, see Asad 1975; Huizer and Mannheim 1979). A similar critique can be made of the early studies of primates and human evolution. This would explain why nonhuman primates and early human populations are often depicted as selfish and aggressive brutes. In this context the discovery by sociobiologists and Marvin Harris that warfare and inequality are basic and universal features of human nature is not surprising.

The 1970s witnessed an introduction of critical new perspectives in anthropology, occasioned by the entrance of more women, African

Americans, other Third World peoples, and students from working-class backgrounds into academia. Whole new avenues of research and theoretical frameworks were developed in the pursuit of antiracist, anti-imperialist, and antisexist theories of human society. The errors of the field's founders, however, cannot be corrected just once. Maintaining integrity requires constant reflection, a continual monitoring of research, and an insistence upon historical specificity. Leacock asserted that, as part of this correction, historical records have to be examined in order to determine whether or not warfare predated contact with European colonialism.

Leacock's Ethnohistory of the Montagnais-Naskapi in Labrador

Leacock's reading of the seventeenth-century writings of the leading Jesuit missionary, Paul Le Jeune documents the degree to which the Montagnais abhorred violence. We learn that "they will not tolerate the chastisement of their children . . . cannot bear to have their children punished, even scolded." When the community rebelled over the decision by the French and Christian Indians to punish a non-Christian Montagnais woman, Le Jeune commented: "Such acts of justice cause no surprise in France. . . . But among these peoples . . . it is a marvel, or rather a miracle, to see . . . any act of severity or justice performed." (quoted in Leacock 1980c:28, 34).

Another example of the cultural distance between the Montagnais and the French colonists on the issue of violence can be found in the story about a French boy who hit a Montagnais man and drew blood. The Montagnais confronted the French authorities, expecting them to respond by giving gifts to make amends, the traditional Montagnais means of making compensation for personal injury. The Montagnais were horrified to learn that, instead, the French planned to flog the boy in public. The Montagnais attempted to intervene, and, Le Jeune reports:

> they began to pray for his pardon, alleging he was only a child, that he had no mind, that he did not know what he was doing; but as our people were nevertheless going to punish him, one of the Savages stripped himself entirely, threw his blanket over the child and cried out to him who was going to do the whipping: "Strike me if thou wilt, but thou shalt not strike him." (quoted in Leacock 1980c, 37)

These events occurred at the moment of initial colonial contact, in the period of immediate confrontation between the gathering/hunting tradition of Montagnais society in Labrador and French state-level society. The stories demonstrate Montagnais shock and outrage over the violent practices of the French colonizers. Again, Le Jeune reports that, "some Savages, having heard that, in France, malefactors are put to death, have often reproached us, saying that we were cruel—that we killed our own countrymen; that we had no sense" (quoted) in Leacock 1980c, 34). For their part, the French were enraged by the Montagnais lack of respect for the structures of hierarchical and patriarchical authority that dominated the colonizers' world. They were frustrated by their inability to make the Montagnais conform to European practices.

Leacock's ethnohistorical studies yielded another corrective for our understanding of culture change. Her research documented that exposure to European practices of violence often preceded actual colonial contact (see, for example, Leacock and Etienne 1980; and Wolf 1982). Indigenous communities situated in coastal and riverine areas experienced the full force of initial European contact. It would be years before the colonizers penetrated into more remote areas. Yet coastal communities were almost always involved in trade, rivalries, and political relations with these other groups of indigenous peoples. By the time the colonizers actually arrived on the scene, these more remote groups had been experiencing the shockwaves of European impact for some time through the old trade networks, including the introduction of new weapons, technology, and European diseases. There was also an increase in intergroup conflict resulting from land pressure, as people from conquered areas moved into more remote zones. Under these circumstances of cultural rupture and dislocation, Europeans claimed that they were observing the authentic or "pristine."

Samoa: Leacock's Final Research Project

Leacock's commitment to careful historical study shaped her final research project: a reexamination of Samoan society. She wanted to critique sociobiologist Derek Freeman's wholesale attack on Margaret Mead's earlier ethnographic work (1983) and she intended to place the contemporary problems of Samoan women and youth in historical perspective (Leacock 1984b, 1986b, and 1987a).

Freeman had lashed out at Mead, claiming that her work was a "myth." According to Freeman, Mead's portrayal of Samoa as a peaceful and sexually relaxed society, in which women enjoyed autonomy and equality with men, was a distorted fantasy. He argued that Mead had ignored the basic genetic structures that program violence and sexual aggression in all human societies, which, to Freeman, are especially prominent in Samoan society.

As a committed sociobiologist, Freeman embarked on his Samoa project with the specific ideological intention of exposing Mead's error and proving that Samoans were really a bunch of violent rapists (Holmes 1987). He focused his work on the more urban port areas. Because of the politics involved, Freeman's book was promoted and reviewed in the media even before it was published. Leacock decided to reexamine the historical literature and study contemporary Samoan society in order to clarify the debate.

Leacock acknowledged that Mead had set off for Samoa with a distorted agenda. She went in search of the "pure" Samoan way of life, untainted and untouched by Western values and contact goods. Mead was simply following one of the core elements of American particularism, the dominant school of anthropology in the United States at the time. Leacock thought that Mead exhibited a profound naïveté and an arrogant disregard for history in her search for some kind of authentic, pristine culture in the most remote areas of western Samoa.

Leacock did not accept Mead's pastoral, romanticized version as a complete picture of Samoan culture. By excluding the more builtup port areas in her analysis, Mead's portrayal was only a partial picture. She had ignored hundreds of years of colonial history, which constitute a significant aspect of Samoan culture. But Leacock agreed that the area studied by Mead had not, in fact, experienced the same degree of cultural dislocation and degradation as the more urban port areas studied by Freeman.

Leacock certainly did not accept Freeman's depiction of Samoan life. She rejected his methodology, especially his uncritical use of nineteenth-century colonial newspaper articles, missionaries' letters, and government records. All of these documents were sensationalist and racist. They were part of the German and British ideological project in Samoa, serving to justify colonialism under the guise of being a "civilizing" influence. Leacock concluded that Freeman's research exhibited a complete lack of attention to the range of social behaviors present in Samoan society. He had made no attempt to understand the correlation between different degrees of colonial penetration and increased violence and social dislocation.

Leacock's historical and ethnographic research in Samoa demonstrated that both Mead and Freeman had allowed larger ideological agendas to distort their anthropological research. In Leacock's opinion, there was much truth in Mead's work, but that truth was specific to the particular areas of Samoa in which she had conducted fieldwork. In similar fashion, Leacock would not deny the presence of poverty and cultural dislocation in the more urbanized port areas of Samoa, problems that reflected the particular economic crisis and social degradation that had resulted from generations of colonial domination.

In line with her commitments to advocacy and applied anthropology, Leacock joined forces with Samoan scholars and community organizers to help develop strategies by which Samoan youth could build programs that would deal with the multitude of social problems they confront. She also worked to support the Samoan people's struggle against imperialism, focusing on structures of economic and cultural dependency and working to expose United States efforts to militarize the entire region. These were the projects she was working on at the time of her death in 1987.

ADVOCACY VERSUS "NEUTRALITY" IN SOCIAL SCIENCE

> I have criticized the anti-historical, functionalist, and psychological emphasis in anthropology that stresses equilibrium, turns attention away from contemporary realities of exploitation and conflict, and downplays possibilities for change (Leacock 1987b, 333)

The AAA discussion on the anthropology of warfare in 1968 was a highly politicized project. Participants insisted that a general, theoretical discussion of war must account for issues of institutionalized power, competing interests, and exploitation. If war serves some kind of "function," the real question is, For *whom*? What underlying interests are being served? (Fried, Harris, and Murphy 1968, 100–101). Not everyone was pleased with this turn of events. Some anthropologists protested the "politicized" discussion. Some went so far as to withdraw their papers from the book in response. They wanted to see an "objective," uncontroversial study of war.

Eleanor Leacock consistently opposed this approach to research. She would ask, How can warfare be neutral? She was fond of a quote by Morton Fried: "When dealing with contending parties, one of which is in power, 'Neutrality' is support of the incumbent" (1972, 45). In her article on theory and ethics in anthropology, Leacock explained that "scientific neutrality is an illusion. . . . Even scientific facts are not 'objective' [but] value-laden" (1987b, 320) She believed that advocacy was crucial for the development of social science research. Leacock argued that "advocacy leads to fuller and more accurate understanding than attempted neutrality" and that "advocacy encourages theoretical innovation to an extent that attempted neutrality does not" (1987b, 323, 333).

In her presentation to graduate students in anthropology at Boston University in 1984, Leacock argued that the most important task facing anthropologists was to develop popular methods of reaching the general public (1984a). She advocated extensive use of the media, as Margaret Mead had done through her column in *Ladies Home Journal* during the 1960s. Every week Mead's ideas entered millions of homes. She stimulated a provocative discussion on complex social problems such as the generation gap, changing sex roles, the counterculture, and the antiwar movement. Without endorsing the particular content of Mead's commentary, Leacock strongly supported this kind of public outreach.

Marvin Harris's books enjoy a wide readership outside academic circles. *Cows, Pigs, Wars and Witches: The Riddle of Culture* (1974), *Cannibals and Kings: The Origins of Culture* (1977), *America Now: The Anthropology of a Changing Culture* (1981), and *Good to Eat: Riddles of Food and Culture* (1985) were all best-sellers. Leacock felt it was important, in this context, to challenge Harris's theories of cultural materialism and expose the tautological reasoning and reductionism inherent in his writings on warfare, violence, and aggression.

Each of Harris's volumes provides compact explanations of social problems, based on a circular reasoning that confuses the consequence with the cause. His books seize upon popular anxieties about changing social relationships and cultural values. These worries are coalesced into provocative titles, including "Why Nothing Works," "Why Women Left Home," "Why the Gays Came out of the Closet," and "Why There's Terror on the Streets" (all are chapter titles in the 1981 volume *America Now*). Changes in the power relations between workers and management, changes in the sexual relations between men and women, and changes in the racial relations between African Americans and European Americans in the United States are presented as a collection of "just-so stories" by Marvin Harris. Complex historical problems are reduced to a set of simple formulas; practices that appear to be bizarre or irrational are described as "functional adaptations." Perhaps his facile explanations account for Harris's popularity; he poses as a "man of science," promising a thorough, probing investigation into the causes of highly charged social problems such as racism, sexism, and homophobia only to deliver simplistic homilies.

In Harris's world the cause of a social problem always seems to reside within the affected population. It is another case of "blaming the victim" and reinforcing the status quo. He pays slight attention to historical developments, and he downplays the differential access to power and resources that exists among social groups. The explanations put forth by Harris form neat packages wrapped in logic and statistics and having no complicated loose ends. His readers can finish a book feeling as though they got "below the surface" to understand the cause of a social problem, while they are allowed to retain a range of unsubstantiated and unreflective ideological notions: that the poor create their poverty; that racism against African Americans is functional for the system; or that war is adaptive and female subordination is necessary.

Harris's discussion of social problems is removed from the realms of human agency and human history, which are replaced by the distant, disinterested, and seemingly objective forces of nature and biology. By accepting Harris's approach, people are simultaneously absolved from their collective responsibility for social problems and denied their generative capacity to transform existing conditions. While some anthropologists have taken issue with Harris (see, for example, Diener, Moore, and Mutaw 1980; and Hirschfeld, Howe, and Levin 1978), most of the opposition has opted simply to dismiss his theory of cultural materialism and ignore his popularity. Eleanor Leacock was determined to challenge Harris, based on her respect for his readership and her concern over the impact of his ideas on United States culture.

CONCLUSIONS

This article has reviewed the anthropological research on warfare in order to discredit the sociobiologist's claims for genetic determinism. The Divale and Harris position that warfare and female infanticide are universal and functionally adaptive has been exposed for what it is: "biological determinism in cultural clothing." And the particular contributions of Eleanor Leacock in developing a more historical and dialectical approach have been outlined.

It is not enough to understand how and why Leacock rejected Harris's and the sociobiologist's approach to the study of war. We also want to consider Leacock's alternative framework for studying the place of women and men in the struggle for peace. While she worked to expose and defeat fraudulent notions of innate male aggressiveness, Leacock also rejected the misguided effort to idealize female passivity and nurturing in order to project the idea that women are, by nature, more peaceable then men. In Leacock's review of Evelyn Reed's *Women's Evolution*, she firmly rejected Reed's projection of a "Matriarchal Sisterhood," arguing that "there is no basis for pitting an image of group-oriented females against one of predatory and cannibalistic males" (Leacock 1981a, 190).

Leacock criticized Reed's attempt to reconstruct a matriarchal narrative of human evolution by "reworking the same stereotyped assumptions about innate feminism and masculine natures which underlie the establishment science Reed purports to contradict." Leacock observed that, because "she does not base her theory of social evolution solidly on the historical materialist framework," Reed constructs a reified image of women's role in human evolution. On the basis of the data from primate studies, Leacock pointed out that "mother-offspring dyads are important in most primate societies, but so are sibling groups, and so is male interest in the young" (1981a, 183, 189).

Summing up her position on "Women as a Force for Peace" at the Second Seneca Falls Women's History Conference, Leacock wrote the following:

> The major part women have taken in building a peace movement in our country is sometimes ascribed to their "female" nature. However, although women are not usually socialized for military activity, cross-cultural evidence does not show them to be inherently more peaceful than men. Instead it demonstrates a close link between militarism and the subordination of women along with slavery and other inequalities. Accordingly . . . women's potential as a force for world peace is better realized when built firmly on their alliance with other oppressed groups rather than on the assumption that as women they are "naturally" peaceful. (1982c)

This passage reflects many of the intellectual, professional, and political concerns that Eleanor Leacock pursued in her career as an anthropologist and throughout her lifetime. She consistently maintained a partisan commitment to the liberation and empowerment of women, nationally oppressed peoples, and the working class. A passion for rigorous historical study strengthened and infused all of her work. Never satisfied to simply challenge and critique racism, sexism, and reactionary social science, Leacock worked vigorously to generate an alternative approach. This is the personal and scientific legacy we inherit from Eleanor Leacock. The search for an authentic liberating science of humanity remains a continuing struggle and our collective responsibility.

NOTES

1. Readers who want to learn more about the multifaceted range and scope of Leacock's scholarship are referred to Christine Ward Gailey's biographical article (1988) and to the Selected Bibliography in this volume.

2. For more information on the issue of ethics in anthropology, see Leacock 1987b.

REFERENCES

Ardrey, Robert. 1961. *African genesis: A personal investigation into the animal origins and nature of man*. New York: Atheneum.

Asad, Talal, ed. 1975. *Anthropology and the colonial encounter*. Atlantic Highlands, N.J.: Humanities Press.

Diener, Paul, Kurt Moore, and Robert Mutaw. 1980. "Meat, markets, and mechanical materialism: The great protein fiasco in anthropology." *Dialectical Anthropology* 5.

Divale, William Tulio (with James Joseph). 1970. *I lived inside the campus revolution*. New York: Cowles Book Company.

Divale, William Tulio and Marvin Harris. 1976. "Population, warfare, and the male supremacist complex." *American Anthropologist* 78:521–38.

______. 1978a. "Reply to Lancaster and Lancaster." *American Anthropologist* 80:117–18.

______. 1978b. "The male supremacist complex: Discovery of a cultural invention. *American Anthropologist* 80:668–71.

Fedigan, Linda Marie. 1986. "The changing role of women in models of human evolution." *Annual Review of Anthropology* 15:25–66.

Freeman, Derek. 1983. *Margaret Mead and Samoa: The making and unmaking of an anthropological myth*. Cambridge, Mass.: Harvard University Press.

Fried, Morton. 1972. *The study of anthropology*. New York: Crowell.

Fried, Morton, Marvin Harris, and Robert Murphy, eds. 1968. *War: The anthropology of armed conflict and aggression*. Garden City, N.Y.: Natural History Press.

Gailey, Christine Ward. 1988. "Eleanor Burke Leacock." In *Women anthropologists: A biographical dictionary*, ed. Ute Gacs, Aisha Kahn, Jerrie McIntyre and Ruth Weinberg, 215–21. Westport, Conn.: Greenwood Press.

Goodale, Jane C. 1971. *Tiwi wives: A study of the women of Melville Island, North Australia*. Seattle: University of Washington Press.

Harris, Marvin. 1968. *The Rise of anthropological theory*. New York: Crowell.

______. 1972. "Warfare old and new." *Natural History* 81(3): 18.

______. 1974. *Cows, pigs, wars and witches: The riddle of culture*. New York: Random House.

______. 1975. *Culture, people, nature: An introduction to general anthropology*, 2d ed. New York: Crowell.
______. 1977. *Cannibals and kings: The origins of cultures*. New York: Random House.
______. 1979. *Cultural materialism: The struggle for a science of culture*. New York: Random House.
______. 1981. *America now: The anthropology of a changing culture*. New York: Simon and Shuster, Touchstone.
______. 1984. "Animal capture and Yanomami warfare: Retrospect and new evidence." *Journal of Anthropological Research* 40:183–201.
______. 1985. *Good to eat: Riddles of food and culture*. New York: Simon and Schuster.
______. 1987a. *The sacred cow and the abominable pig: Riddles of food and culture*. New York: Simon and Schuster.
______. 1987b. *Why things don't work: The anthropology of daily life*. New York: Simon and Schuster. (Reprinted from *America Now* [1981].)
______. 1987c. "Foodways: Historical overview and theoretical prolegomenon." In *Food and evolution: Toward a theory of human food habits*, ed. Marvin Harris and Eric B. Ross, 57–90. Philadelphia: Temple University Press.
______. 1989. *Our kind: Who we are, where we came from, where we are going*. New York: Harper and Row.
Harris, Marvin, and Eric B. Ross, eds. 1987a. *Food and evolution: Toward a theory of human food habits*. Philadelphia: Temple University Press.
______. 1987b. *Death, sex, and fertility: Population regulation in pre-industrial and developing societies*. New York: Columbia University Press.
Hirschfeld, Lawrence A., James Howe, and Bruce Levin. 1978. "Warfare, infanticide, and statistical inference: A comment on Divale and Harris." *American Anthropologist* 80:110–15.
Holmes, Lowell, ed. 1987. *Quest for the real Samoa: Assessing the Mead/Freeman controversy and beyond*. Granby, Mass.: Bergin.
Howe, James. 1978. "Ninety-two mythical populations: A reply to Divale et al." *American Anthropologist* 80:671–73.
Huizer, Gerrit, and Bruce Mannheim, eds. 1979. *The politics of anthropology: From Colonialism and sexism toward a view from below*. The Hague: Mouton Publishers.
Hymes, Dell, ed. 1974. *Reinventing anthropology*. New York: Vintage.
Lancaster, Chet, and Jane B. Lancaster. 1978. "On the male supremacist complex: A reply to Divale and Harris." *American Anthropologist* 80:115–17.
Lancaster, Jane B. 1973. "In praise of the achieving female monkey." *Psychology Today*, September.
Leacock, Eleanor. See the separate bibliographic sec. in this volume.
Lee, Richard B. 1979. *The !Kung San: Men, women, and work in a foraging society*. Cambridge and New York: Cambridge University Press.

Leibowitz, Lila. 1979. " 'Universals' and male dominance among primates: A critical examination." In *Genes and gender II: Pitfalls in research on sex and gender*, ed. Ethel Tobach and Betty Rosoff, 35–48. New York: Gordian Press.

Morgan, Lewis Henry. [1877] 1963. *Ancient society*. New York: Meridian.

Morris, Desmond. 1967. *The naked ape: A zoologist's study of the human animal*. New York: McGraw-Hill.

Nash, June. 1978. "The Aztecs and the ideology of male dominance." *Signs* 4.

Norton, Helen H. 1976. "The male supremacist complex: Discovery or invention?" *American Anthropologist* 78:665–67.

Rohrlich-Leavitt, Ruby. 1975. *Peaceable primates and gentle people: Anthropological approaches to women's studies*. Harper Studies in Language and Literature Module. New York: Harper and Row.

______. 1977. "Women in transition: Crete and Sumer." In *Becoming visible: Women in European history*, ed. Renate Bridenthal and Claudia Koonz. Boston: Houghton Mifflin.

Sider, Gerald, ed. 1974. *The shaping of American anthropology, 1883–1911: A Franz Boas reader*. New York: Basic Books.

Slocum, Sally. 1975. "Woman the gatherer: Male bias in anthropology." In *Toward an anthropology of women*, ed. Rayna R. Reiter. New York: Monthly Review Press.

Tanner, Nancy, and Adrienne Zihlman. 1976. "Women in evolution, Part 1: Innovation and selection in human origins." *Signs* 1.

Trinkaus, Erik. 1978. "Hard times among the Neanderthals." *Natural History* 87.

Wolf, Eric R. 1982. *Europe and the people without history*. Berkeley: University of California Press.

Wolf, Eric, and Joseph Jorgensen. 1970. "Anthropology of the warpath in Thailand." *New York Review of Books*, 19, November, 26–35.

Zihlman, Adrienne, and Nancy Tanner. 1978. "Gathering and the hominid adaptation." In *Female Hierarchies*, ed. L. Tiger and H. T. Fowler, 163–94. Chicago: Beresford.

ELEANOR BURKE LEACOCK (1922–1987)

I. SELECTED WRITINGS

1949 — "The Seabird Community." In *Indians of the Urban Northwest*, ed. Marian Smith. New York: Columbia University Press.

1955 — "Matrilocality in a Simple Hunting Economy (Montagnais-Naskapi)." *Southwestern Journal of Anthropology* 11:31–47.

1958 "Status among the Montagnais-Naskapi of Labrador." *Ethnohistory* 5:200–09.

1963 Introduction to *Ancient Society*, ed Lewis Henry Morgan. 1877. Reprint. New York: Meridian.

1967 "Distortions of Working-Class Reality in American Social Science." *Science and Society* 31(1): 1–21.

1968 "Personality and Culture Theory in the Field of Education." In Vol. 2, *Proceedings of the 8th International Congress of Anthropological and Ethnological Sciences*. Tokyo: Science Council of Japan.

1969a *Teaching and Learning in City Schools: A Comparative Study*. New York: Basic Books.

1969b "The Montagnais-Naskapi Band." In *Contributions of Anthropology: Band Societies*, ed. David Damas. National Museum of Canada Bulletin 228. Ottawa: Queens Printer for Canada.

1970 "Education, Socialization, and the 'Culture of Poverty.'" In *Schools against Children: The Case for Community Control*, ed. Annette T. Rubenstein. New York: Monthly Review Press.

1971a *Culture of Poverty: A Critique*. New York: Simon and Schuster.

1971b "Theoretical and Methodological Problems in the Study of Schools." In *Anthropological Perspectives on Education*, ed. Murray Wax. New York: Basic Books.

1972a Introduction to *The Origin of the Family, Private Property, and the State*, by Friedrich Engels. New York: International Publishers.

1972b "Abstract vs. Concrete Speech: A False Dichotomy." In *Functions of Language in the Classroom*, ed. Courtney Cazden et al. New York: Teacher's College Press.

1972c "At Play in African Villages." *Natural History* 80 (December): 60–65.

1973 *Primary Schooling in Zambia*. Bethesda, Md.: ERIC Document Reproduction Service.

1974 "Learning, African Style." In *Notes from Workshop Center for Open Education*. New York: City College Workshop Center for Open Education.

1975 "Class, Commodity, and the Status of Women." In *Women Cross-Culturally: Change and Challenge*, ed. Ruby Rohrlich-Leavitt. The Hague: Mouton.

1976a Review of *Marxist Analyses and Social Anthropology*, ed. Maurice Bloch. *Contemporary Sociology* 5:671–720.

1976b "*The Montagnais Hunting Territory and the Fur Trade.*" *American Anthropologist*. Memoir 78:759.

1977a "Education in Africa: Myths of Modernization." In *The Anthropological Study of Education*, ed. Francis Ianni and Craig Calhoun. The Hague: Mouton.

1977b "Race and the We-They Dichotomy in Culture and Classroom." *Anthropology and Education Quarterly* 8(7).

1977c "The Changing Family and Levi-Strauss, or Whatever Happened to Fathers?" *Social Research* 44(2): 235–59.

1977d "Women in Egalitarian Society." In *Becoming Visible: Women in European History*, ed. R. Bridenthal and C. Koonz. Boston: Houghton Mifflin.

1978a. "Society and Gender." In *Genes and Gender*, ed. Ethel Tobach and Betty Rosoff. Genes and Gender, vol. 1. New York: Gordian Press.

1978b "Women's Status in Egalitarian Society: Implications for Social Evolution." *Current Anthropology* 19(2): 24–75. With Commentary by Ronald Chen, Bruce Cox, and Ruth Landes.

1979 "Women, Development, and Anthropological Facts and Fiction." In *The Politics of Anthropology: From Colonialism and Sexism toward a View from Below*, ed. Gerrit Huizer and Bruce Mannheim. The Hague: Mouton.

1980a "Politics, Theory, and Racism in the Study of Black Children." In *Theory and Action: Essays to Gene Weltfish*, ed. Stanley Diamond. The Hague: Mouton.

1980b "Social Behavior, Biology and the Double Standard." In *Sociobiology: Beyond Nature/Nature?*, ed. George Barlow and Allen Silverberg. Boulder, Colo.: Westview Press.

1980c "Montagnais Women and the Jesuit Program for Colonization." In *Women and Colonization: Anthropological Perspectives*, ed. Mona Etienne and Eleanor Leacock. New York: Praeger.

1981a *Myths of Male Dominance: Collected Articles on Women Cross-Culturally*. New York: Monthly Review Press.

1981b "Seventeenth-Century Montagnais Social Relations and Values." In *Subarctic*, ed. June Helm. Handbook of North American Indians 6(3): 190–95. Washington: Smithsonian Institution.

1981c "History, Development and the Division of Labor by Sex: Implications for Organization." *Signs* 7(2):471–91.

1982a "Relations of Production in Band Society." In *Politics and History in Band Societies*, ed. Eleanor Leacock and Richard B. Lee. New York: Cambridge University Press.

1982b "Marxism and Anthropology." In *The Left Academy: Marxist Scholars on American Campuses*. ed. Bertell Ollman and Edward Vernoff. New York: McGraw-Hill.

1982c "Women as a Force for Peace." Abstract for a paper presented as part of the panel "Anthropological Perspective on Women and Community." Second Seneca Falls Women's History Conference, 17 July 1982.

1983 "Interpreting the Origins of Gender Inequality: Conceptual and Historical Problems." *Dialectical Anthropology* 7:263–84.

1984a "Integrating the Personal and Political in Anthropology." Presentation to the Graduate Student Seminar, Department of Anthropology, Boston University, Boston, Mass.

1984b "Compounding the Problems of Samoan Youth." *Kaleidoscope* 1 (2). New York: The City College.

1986a "The Anthropology of War: Biological Determinism in Cultural Clothing." Notes and outline for a paper presented at the Genes and Gender Conference, Stern College, New York, N.Y.

1986b "Adolescents et relations entre les sexes aux Samoa de nos jours: à propos des accusations de Derek Freeman contre Margaret Mead." Lecture delivered at the College of France.

1987a "Postscript: The Problems of Youth in Contemporary Samoa." In *Quest for the Real Samoa: Assessing the Mead/Freeman Controversy and Beyond*, ed. Lowell Holmes. Granby, Mass.: Bergin and Garvey.

1987b "Theory and Ethics in Applied Urban Anthropology." In *Cities of the United States: Studies in Urban Anthropology*, ed. Leith Mullings. New York: Columbia University Press.

II. COAUTHORED WORKS

Leacock, Eleanor, Martin Deutsch, and Joshua A. Fishman.

1965 *Toward Integration in Suburban Housing: The Bridge-View Study*. New Jersey: Anti-Defamation League of B'nai B'rith.

Leacock, Eleanor, L. Menashe et al.

1969 *Social Science Theory and Method: An Integrated Historical Introduction*. Vols. 1–8. New York: Polytechnic Institute of Brooklyn.

Leacock, Eleanor, and Jacqueline Goodman.
1976 "Montagnais Marriage and the Jesuits in the Seventeenth Century: Incidents from the Relations of Paul Le Jeune." *Western Canadian Journal of Anthropology* 6(3): 77–91.

Leacock, Eleanor, and June Nash.
1977 "Ideologies of Sex: Archetypes and Stereotypes." In *Issues in Cross-Cultural Research. The Annals of the New York Academy of Sciences* 285. New York: New York Academy of Sciences.

Leacock, Eleanor, and Vera John.
1979 "Transforming the Structure of Failure." In *Educating All Our Children: An Imperative for Democracy*, ed. Doxey Wilkerson. Westport, Conn.: Mediax.

Leacock, Eleanor, and Mona Etienne, eds.
1980 *Women and Colonization: Anthropological Perspectives*. New York: Praeger.

Leacock, Eleanor, and Edward S. Rogers.
1981 "Montagnais-Naskapi." In *Subarctic*, ed June Helm. Handbook of North American Indians 6(3): 169–89.

Leacock, Eleanor, and Richard B. Lee, eds.
1982 *Politics and History in Band Societies*. New York: Cambridge University Press.

Leacock, Eleanor, and Helen I. Safa, eds.
1986 *Women's Work: Development and the Division of Labor by Gender*. 2d ed. Granby, Mass.: Bergin and Garvey.

Leacock, Eleanor, and Nancy O. Lurie, eds.
1988 *North American Indians in Historical Perspective*. Prospect Heights, Ill.: Waveland Press.

♦ GEORGINE SANDERS ♦

War, Then and Now

Thunder in the cloudless dawn
low flying planes, so came the war
on that first perfect day of spring.
How soon we learned its bitter words

of pain, fear, death and loss of love,
all part of twisted daily life,
of bravery and sacrifice
and every kind of treason.

We hungered for the common things.
The shrinking world bore down upon
our streets, where soldiers marched in song
and captives marched toward their doom.

I tried throughout the years of peace
to lose those fateful words in vain.
They still describe reality:
war has many other names.

♦ I ♦

The Misuse of Scientific Research: Examples in Genetics, Endrocrinology, and Biopsychology

♦ REGINA E. WILLIAMS ♦

Mothermilk

from birth
i've waited
for breasts and mothermilk
to touch all unexpressed wants
and whims

waited
for fingertip dreams
and fantasies to be forged
by mother and me
waited for momma's fist
to fight all 4-headed dragons
and 2-legged fears
waited to share secrets of
herbs and hairstyles
of what to do with tampons
waited for her stroke to touch
my undefined pain before it surfaced
into shout
waited
for a taste of mothermilk

but a war is on
ammunition is on order

and mother has enlisted
to help silence the enemy

so i've waited
and "The Shadow" and a book
became my companions

through matrimonial murders
i've waited
for a stroke to sooth
my unmouthed scream
waited for someone to know
texture of my pain
before it reached surface
to hold me close to the nipple
of love

but the war goes on
and touch is reserved
for my giving:
to little ones with
natural yearnings
to older ones with
demanding wants
to strangers waving flags
of truce

from birth to graying hair
i've waited
felled somewhere on the battlefield
without knowing touch
or taste of mothermilk

    do i have to die
    for this war to end?

i will offer my breasts (in truce)
but i'll never surrender
til i know the taste of mothermilk

♦ BETTY ROSOFF ♦

Genes, Hormones, and War

The purpose of this essay is to examine the genetic and hormonal explanations for war and peace and to determine whether science has been misused in attempts to support these concepts. Science can be defined as the body of facts, laws, and theories concerning real phenomena that the social institutions of science, using the methods of science, claim to be true. Therefore, there are two aspects that must be considered in describing science. First, is the description accurate? Does it correspond to the real world, and can it be independently verified? Second, does the description conform to and support the ideals of the society in which the scientist is working so that research will be supported financially and through publications? What passes as "acceptable" science at any historical moment is determined by the society and has a function in that society (Lewontin 1984). Thus, "bad science" may be based on poor and inaccurate methodology, or it may be based on an ideology that asks the wrong questions and gives benefits to a few while harming the many. In a sexist society in which war is an instrument of national policy, for example, research attempting to establish a genetically determined role for men in war making would be supported and encouraged.

Among well-known examples of bad science are the experiments of the Nazi scientists in support of their "racial theories"; the studies of Burt and Jensen attempting to show that "intelligence" is inherited (Lewontin 1984); the sociobiology of Wilson, which claims that, since genes determine social relationships and the social order is a product of evolutionary adaptation, nothing can be done about wars, hunger, homelessness, sexism, and racism.

One additional problem that makes it more difficult still to distinguish between "good science" and the abuse of science is the notion that science is "pure." This idealization of science is promoted particularly by the media. Often, science and scientists are presented as "objective," and the publishing of fraudulent data is considered a sin perpetrated against the profession. Science is used often by the press as a neutral source of information and a powerful source of authority for the creation of social policy and the support of popular beliefs (Nelkin 1987). Journalists regularly publish preliminary reports of research that have not yet been verified by other scientists, promoting the idea that, if information comes from a scientist, it is fact.

With this idealization of science in mind, we must examine the research on genes, hormones, and war. During the past thirty years we have made tremendous progress in understanding the gene, but the most common misuse of that knowledge is to attribute to genes powers to determine all aspects of life, human behavior, even the formation of human society.

Genetic determinists have something to say about war and peace as they do about every societal and political question. The ultimate tenet of their argument is that the force motivating animal behavior is the passing on of genes to the next generation. In order to compete effectively with others, therefore, there are genes for aggression found in the male. This competition occurs in humans, as it does in animals and explains human evolution. According to genetic determinists, since men are genetically programmed for aggression, fighting in wars is "natural."

But in fact there is a great deal of difference between personal, individual aggression and war. Abnormal physiological processes (due to various emotional factors) might lead to the killing of one person by another. We label behavior psychotic when one individual kills several people whom he or she doesn't know. But we don't say that killing many people is psychotic when we are at war. In order for men to fight in a war, some rethinking of normal behavior must be involved. If people are needed by those in power to fight their brothers and sisters in another country, then racism is used to convince them to do it. If people believe that they must overthrow their own oppressors, then they are persuaded that a revolutionary war is their only choice.

Some genetic determinists who promote peace propose that, since men have the genes for aggression, we should put women into government because their genes make them peaceful and nurturing (Konner 1982). Yet, genetic determinists fail to explain the following:

1. Men in recent history have to be drafted to fight in a war.
2. Men refuse to fight in an "unjust war" but will fight in a "just war."
3. Women in the course of history have been resistance and guerrilla fighters for causes they believe in.
4. In an egalitarian society wars will not be necessary to support the economy.

Sociobiology, the current theory of genetic determinism, is based on the idea that behavior patterns can be traced directly to our genes. Lumsden and Wilson (1981) propose to call the genes involved in societal behavior "culturgens." War and the means of making war, they maintain, are controlled by culturgens. Generalizing from genes to complex behavior patterns, however, ignores the real facts of biology. Genes are pieces of DNA, a chemical that contains the information to make many different proteins in the environment of the cell. These proteins make up a large part of body structure, and they also determine the functioning of different tissues and organs by catalyzing chemical reactions in the cells. A skin cell does not make hemoglobin, and a red blood cell does not synthesize keratin. All the cells of any one individual, however, contain the same genes; therefore, the cellular environment must play a role in determining what genes are "turned on" to make a particular protein. For example, the genes for making the protein that causes muscle contraction are found in all body cells, but only in the environment of the muscle cell is it turned on so that muscles contract.

If we look at endocrine cells that make hormones, we find that some hormones are proteins, such as the growth hormone made in the pituitary gland. Other hormones are made by further processing proteins. For example, the thyroid hormone is made by adding iodine to proteins made in the thyroid gland. The sex hormones, estrogens and androgens, are made from cholesterol using as catalysts proteins produced by the genes. The hormones are then transported to cells that have the receptor for that particular hormone, and a specific action is produced. During puberty, for example, androgens stimulate the growth of muscle cells in boys, while estrogens cause the growth of breast cells in girls. Generally, individual cells do not act independently but combine into tissues, organs, and systems within the body. We must look at all of these levels within the organism, therefore, to see how they interact to produce a particular effect. But the consequences of all these physiological processes taking place within the endocrine and nervous systems are integrated into behavior conditioned by an individual's experience in society. Tracing the cause of war to the genes that make the proteins that in turn catalyze the synthesis of androgens (the "aggression" hormones) does not make much biological sense.

The principle androgen is testosterone, the hormone most often identified with aggressive behavior. This hormone is produced in the testes in men and the adrenal glands in both sexes and is transported in the blood to its target organs. Testosterone clearly plays an important role in the development and maintenance of the male reproductive system and some of the secondary sexual characteristics like hair growth and muscle development. Behavior involved in the reproductive process seems to be influenced by testosterone in male animals and to a lesser extent in humans (Martin 1985). But what about nonreproductive behavior like aggression? Is it also influenced by testosterone?

Several different types of studies have attempted to show that aggressive behavior and hormonal levels of testosterone are correlated. At least three methodological questions are critical to all of these studies:

1. What is aggressive behavior, and how is it measured? In psychological tests used for the studies, hostility, ambition, depression, and dominance are all equated with aggression. Some other studies are observations of "tomboy," or athletic, behavior in children as indications of aggressiveness. Some psychologists use prisoners as subjects, suggesting that criminality is an example of aggressive behavior. Other methods used involve reports by family members of subjects' aggression as well as their reactions to simulated situations (Fausto-Sterling 1985). The possible bias of the tests and the observers in our male-dominated society is not reported nor even mentioned.

2. Most of these studies are based on a single blood determination of testosterone. Human and other primate studies show great variation in blood levels of testosterone at different times in a single day, however, and some experiments even suggest cyclical patterns over several days (Briscoe 1978). In one study of vervet monkeys, the levels of testosterone varied tenfold in one day (Steklis et al. 1985). Thus, the variations in one individual in one day might be greater than the differences between aggressive and nonaggressive groups when only one blood sample is taken.

3. Physiologically, the endocrine system functions through a process in which many hormones act together to produce a particular effect. In order to maintain a normal blood sugar, for example, insulin is secreted after a meal to reduce the blood sugar, while glucagon and cortisone raise blood sugar between meals. In addition, many hor-

mones have the same effect on an organism but function through different mechanisms. An individual with insufficient thyroid or adrenal hormones is far more apathetic than one lacking only testosterone (Martin 1988). When an individual feels stress and behaves aggressively, the actions are accompanied by secretion of large quantities of adrenal hormones (Martin 1985). A number of studies have shown that testosterone is actually reduced under stress (Martin 1985; Bleier 1984).

In studies assessing the relationship between aggressive feelings or behavior and blood levels of testosterone, there does not seem to be any clear-cut correlation between high testosterone levels and aggression. When twenty-four college men were tested, one-half scored high and the other half scored low on hostility tests, and there was no correlation with their blood testosterone levels (Fausto-Sterling 1985). In another test of twenty healthy volunteers, there was no consistency between hormone levels and high scores for aggression or hostility (Fausto-Sterling 1985).

A recent report of two studies that were performed by the same group purported to measure relationships between aggressive behavior and hormone levels in young adolescents (9–14 years of age). The first study of fifty-six boys and fifty-two girls, where aggression was measured by the reporting of mothers, showed that higher levels of another androgen, *androstenedione* correlated with reported aggressive behavior in boys but not in girls[1] (Susman et al. 1987). In the second study of boys and girls in the same age group, aggressive behavior was measured by evaluating videotaped family interactions during problem-solving tasks (Inhoff-Germain et al. 1988). In this study, the higher levels of androstenedione were associated with aggressive behavior in girls but not boys. The contradictory results support the thesis that hormones and aggressive behavior are not clearly related.

In two reports involving prisoners, there was no clear correlation between psychological test scores for aggression and testosterone levels. In one study of prisoners there was a suggestion that aggressive behavior might be correlated with high hormone levels, but not in the other (Fausto-Sterling 1985). In fact, even in cases of sexual aggression, testosterone levels might not be increased. In a recent study of abnormally sexually aggressive men there was no significant difference in their testosterone levels when compared with normal controls (Langevin et al. 1988; Christiansen and Knussman 1987).

In these studies two facts must be considered. Where there might be a correlation between aggressive behavior and increased testosterone levels, the increased testosterone could be the result and not the cause of aggressive actions or feelings (Bleier 1984). In fact, competitive sports and combat in humans have been shown to raise testosterone levels (Piacente 1986). On the other hand, some experimental results have shown that stress lowers testosterone levels, and aggression surely is associated with stress (Martin 1988).

Sociobiologists emphasize the role of hormones in development as an important influence on adult behavior. For example, Barash writes, "Girls accidentally exposed to testosterone while still in their mother's uterus were found to be masculinized as children. They often developed into tomboys and were generally more active than normal girls." He then draws the conclusion that "there are real differences between little boys and little girls, behavioral differences that begin early in life and derive at least in part from our biology" (1979).

Some of the studies often cited to show the effects of prenatal hormones on postnatal behavior development are those in which mothers were given low doses of various synthetic progestins during pregnancy. Progestins have structures like the hormone progesterone and often change to testosterone inside the body. Reinisch administered tests that were supposed to measure the potential for aggressive behavior in both boys and girls from progestin-treated mothers (1981). The author found that both scored higher than their siblings. Ehrhardt and Meyer-Bahlburg, on the other hand, reported that the boys born to progestin-treated mothers did not differ significantly from a control group of boys in terms of aggressive behavior (1981). The girls born to progestin-treated mothers were less inclined to label themselves tomboys and showed a preference for wearing female clothing and playing with dolls.

The contradictory results certainly raise serious questions. First, some progestin compounds change to androgens and others to estrogens in the body, although no distinction was made in the studies between the various types. Another consideration is the fact that during pregnancy normal levels of progesterone rise as high as three hundred milligrams, while the dose given to the mothers in these studies was twenty-five milligrams. It is hard to imagine that so small a dose would have an additional significant effect. These hormones are administered to mothers for a variety of problems during pregnancy. The mothers' own stress, which causes the secretion of other hormones, might have greater effect on the developing fetus than the administered progestins. For example, the sibling control group in the Reinisch study had not been born after problem pregnancies, a factor that was not controlled for.

Another aspect of the studies that should be considered is that androgens are anabolic and therefore increase muscle growth. Thus, children exposed to testosterone might be bigger and stronger than others and therefore better at athletics, which is then interpreted for girls as tomboy behavior.[2] Certainly these reports do not present clear evidence of the effect of testosterone or any androgens on aggressive behavior during development.

Another group of studies, purporting to show the primary effect of male hormones on behavior, were performed on children with andreno-genital syndrome (AGS). These children were genetically female (XX) but were born with male external genitalia. AGS is a disease in which the adrenal gland secretes large amounts of testosterone. During embryonic development it is testosterone that causes the formation of the scrotum, the penis, and other male accessory reproductive organs. The children in these studies had corrective surgery from one to seven years after birth so that their external genitalia appeared to be female. Retrospective studies based on interviews and observations of the children and their families caused the researcher to conclude that the children showed more tomboy behavior than their siblings, although they did not necessarily initiate fighting more often (Money and Ehrhardt 1972; Ehrhardt and Baker 1974). There are many flaws in these studies, a few of which are mentioned below (for further analysis, see Bleier 1984; and Fausto-Sterling 1985):

1. The studies fail to consider the effects on these young children of having male genitalia before surgery was performed, nor do they report on the attitude of family members and peers toward the child. When a child looks anatomically male, for example, the child is expected to be more physically active and to fight more.

2. The AGS children in the study were given daily doses of cortisone, though the family control subjects did not receive this adrenal hormone. Cortisone is one of the hormones normally released during stress; it causes hyperactivity and mood elevation.

3. The choice of behavior types observable for the study was sexist: active play and fighting were associated with masculine behavior; and grooming and parenting with feminine behavior (Gordon 1983).

4. Data collection was biased in that both the parents and the observers knew that these children were female and had—previously or presently—male external genitalia.

Unfortunately, these studies have been widely quoted and used to support the thesis that the fetal brain is changed in some way by androgens to produce aggressive behavior.

Chromosome studies also fail to support a genetic basis for aggressive behavior. In humans, the possession of a Y chromosome in most cases signifies a person of male sex (Probber and Ehrman 1978). Sometimes individuals have two Y chromosomes (that is, XYY instead of the usual XY). Given the assumption that men are more aggressive than women, having two Y chromosomes should make someone even more aggressive. In 1967 a small study of a prison population reported a higher percentage of men in prison than in the general population had two Y chromosomes (Price and Whatmore 1987). The authors of the study concluded that the gene for aggression lies on the Y chromosome and that two Y chromosomes increase the likelihood of becoming a criminal. Widely publicized, this study was used to support the myth of a genetic basis for crime. Accepting such a view, one would not have to examine such societal causes of crime as poverty and oppression, since the true cause lies in the genes. Further analysis of this study and several subsequent studies eventually proved that XYY individuals were not more numerous in the prison population than in the general population and that the crime they had committed was robbery, not murder (Woodward 1981). Psychological tests given to XYY men did not show that they were more aggressive than the normal XY men. In addition, measurements of testosterone levels in the blood of XYY men did not show more of the hormone than that found in a group of normal XY men (Witkin et al. 1977).

Attempts to show behavioral patterns based on genes in the female are equally inconclusive. Women who have three X chromosomes are not more placid than those with the normal XX complement, nor are their estrogen levels different. Even though the myth of the aggressive gene on the Y chromosome has been disproved, the popular press and some textbooks still present the notion as fact.

If hormones and genes cannot explain why wars occur, perhaps the evolution of the human brain can give us some answers to the question. In the course of human evolution, the primate brain has become greatly specialized and complex. This development is evident in the behavioral plasticity of humans—for example, in the ability of each individual to learn a wide range of actions to fit different situations without a single stereotyped pattern. This complexity is also evident in the capacity of humans to formulate problems and solve them. Other species became extinct because they could not change the environment to prevent their destruction. The survival and growth of humans have been possible because of their more complex and plastic brain.

This greater cognitive advantage of humans is not the result of genetic or hormonal processes. Humans can and do utilize knowledge of the past gained through cultural and societal processes rather than depend for survival on their own life experiences. Scientists cannot equate the human brain with the animal brain, as sociobiologists and ethologists do (Tobach 1986).

Attempts of genetic determinists to show a biological basis for individual aggression and to link this to social aggression, are not only unscientific, but they support the idea that wars of conquest between nations are inevitable. These wars are conceived and executed by women and men in governments that serve the interests of the military-industrial complex which makes money out of war. Most men who have to fight in wars do not want to, because they do not want to die themselves, and they do not want to kill other people for the benefit of the war makers on both sides. Such working men, students, farmers, and professionals must work together with women to change this profit-making system and achieve peace. Those who make money out of war work together and sell arms to one another. Others, the victims of war, must not be divided by nationality, color, gender, or religion. The sociobiologists explain racism by saying that our genes program us to defend and fight for those who have more genes in common with us in order to perpetuate our genes. This rationale is another pseudoscientific argument used to divide people and get them to fight wars of national oppression.

As this essay indicates, science is not neutral, for it can be misused against the interests of people. To achieve a science that works for people, scientists and journalists must effectively educate and communicate with the community at large so that informed decisions can be made. Scientists must ask themselves the question, "How can my scientific skills best be used to serve people, to expose and correct the role of science and technology in wreaking genocide in wars, in oppressing individuals and minorities, and in permitting malnutrition and disease throughout the world?" (Rose and Rose 1973). The way to peace is for men and women, scientists and nonscientists, to work together to achieve a peaceful, egalitarian society.

NOTES

1. Androstenedione is closely related to testosterone in structure and function. It is found in both sexes in pre- and early adolescence and acts like a weak androgen.

2. Tomboy behavior is used to loosely describe the behavior of a girl who is physically active, interested in athletics and rough-and-tumble play, and seems to prefer not to play with dolls.

REFERENCES

Barash, D. 1979. *The whispering within*. New York: Harper and Row.

Bleier, R. 1984. *Science and gender*. New York: Pergamon Press.

Briscoe, A. 1978. Hormones and gender. In *Genes and gender*, edited by E. Tobach and B. Rosoff. Genes and gender, vol. 1. New York: Gordian Press.

Christiansen, K., and R. Knussman. 1987. Androgen levels and components of aggressive behavior in men. *Hormones and Behavior* 21:170–80.

Ehrhardt, A. A., and S. W. Baker. 1974. Fetal androgens, human central nervous system differentiation, and behavior sex differences. In *Sex differences in behavior*, edited by R. C. Friedman, R. M. Richart, and R. L. Vandewiele. New York: Wiley.

Ehrhardt, A. A., and H. F. L. Meyer-Bahlburg. 1981. Effects of prenatal sex hormones on gender-related behavior. *Science* 211:1312–18.

Fausto-Sterling, A. 1985. *Myths of gender*. New York: Basic Books.

Gordon, S. 1983. What's new in endocrinology? Target: Sex hormones. In *The second X and women's health*, edited by Myra Fooden et al. Genes and gender, vol. 6. New York: Gordian Press.

Inhoff-Germain, G., G. S. Arnold, E. D. Nottelmann, and E. J. Susman. 1988. Relations between hormone levels and observational measures of aggressive behavior of young adolescents in family interactions. *Development Psychology* 24(1):129–39.

Konner, M. 1982. *The tangled wing: Biological constraints on the human spirit*. New York: Holt, Rinehard and Winston.

Langevin, R., J. Bain, G. Wortzman, S. Hucker, R. Dickey, and P. Wright. 1988. Sexual sadism: Brain, blood, and behavior. In *Human sexual aggression*, edited by R. Prentky and V. Quinsey. *Annals of the New York Academy of Sciences* 528.

Lewontin, R. C., S. Rose, and L. J. Kamin. 1984. *Not in our genes*. New York: Pantheon.

Lumsden, C. J., and E. O. Wilson. 1981. *Genes, mind and culture: The coevolutionary process*. Cambridge, Mass.: Harvard University Press.

Martin, C. 1985. *Endocrine physiology*. New York: Oxford University Press.

Martin, C. 1988. Brain differences: Facts and myths. In *Women at Work: Socialization toward inequality*, edited by D. Burnham, S. Gordon, and G. Vroman. Genes and gender, vol. 5. New York: Gordian Press.

Nelkin, D. 1987. *Selling science*. New York: W. H. Freeman.

Money, J., and A. A. Ehrhardt. 1972. *Man and woman, boy and girl: The differentiation and dimorphism of gender identity from conception to maturity*. Baltimore, Md.: Johns Hopkins University Press.

Piacente, G. J. 1986. Aggression. *Psychiatric Clinics of North America* 9(2):329–39.

Price, W., and P. Whatmore. 1967. Criminal behavior and the XYY male. *Nature* 213:815–16.

Probber, G., and L. Ehrman. 1978. Pertinent genetics for understanding gender. In *Genes and gender*, edited by E. Tobach and B. Rosoff. Genes and gender, vol. 1. New York: Gordian Press.

Reinisch, J. 1981. Prenatal exposure to synthetic progestins increase potential for aggression in humans. *Science* 211:1171–73.

Rose, S., and H. Rose. 1973. Can science be neutral? *Perspectives in Biology and Medicine* 16(4):605–24.

Steklis, H. D., G. L. Brammer, M. J. Raleigh, and M. T. McGuire. 1985. Serum testosterone, male dominance, and aggression in captive groups of vervet monkeys. *Hormones and Behavior* 19:154–63.

Susman, E. J., G. Inhoff-Germain, E. D. Nottleman, and D. L. Loriaux. 1987. Hormones, emotional dispositions and aggressive attributes in young adolescents. *Child Development* 58(4):1114–34.

Tobach, E. 1986. Evolutionary theories and the issue of nuclear war: Implications for mental health. *International Journal of Mental Health* 15:56–64.

Witkin, H. A., S. Mednick, F. Schulsinger, E. Bakkestrom, K. O. Christiansen, D. R. Goodenough, K. Rubin, and M. Stocking. 1977. Criminality in XYY and XXY men. *Science* 193:545–55.

Woodward, V. 1981. *Heredity and human society*. Minneapolis, Minn.: Burgess.

♦ SUZANNE R. SUNDAY ♦

Biological Theories of Animal Aggression

It might seem strange to contribute a discussion on aggression in animals to a book on peace and war, but the majority of researchers who suggest that there are genetic and physiological factors leading to aggressive men and passive women draw heavily on the nonhuman data concerning aggression. Therefore, to properly criticize those who view human aggression as being biologically determined, it is necessary to examine the data on physiological and genetic factors and aggression in nonhuman animals.

In the 1940s there was an expansion of research on inborn, innate, or instinctive sources of aggression with respect to genetics, hormones, and the brain. The majority of this early work centered on mice but examined rats, cats, and primates as well. It was also widely accepted that aggressive behavior was more prevalent in males than females in a variety of species—particularly among mammals. It was precisely this "universal" sex difference that led researchers to presuppose a "biological" basis of aggression and explore the possible genetic and hormonal controls of aggression.

Field and laboratory observations seemed to support the idea of sex-based difference. Different levels of aggression of males in a variety of species (for example, birds, rodents, and some primates) appeared to be associated with seasonal variations, and the seasonal changes were correlated with hormonal changes. In a number of species, aggressive behavior appeared first at puberty, coinciding with the male's large increase in hormone levels. Juvenile male mice, which usually show low levels of aggression, demonstrate much higher levels of aggression after hormone (testosterone) treatment. Further, it was found in many species that castration decreased the male's aggression and hormone replacement reinstated it (see Leshner 1978).

This essay reviews two approaches based on the concept that there is some relationship between human and nonhuman aggression: that of the physiological psychologist (including behavioral endocrinology and behavior genetics) and that of the sociobiologist. Although they are somewhat different approaches, both are based on the genetic determinism of behavior. Their integration has produced much of the current basic theory of animal aggression, with the later approach focusing on the so-called ultimate, or evolutionary, causes of behavior and the former on the "proximate" causes, that is, that biological mechanisms control behaviors.

Few researchers have assumed that humans behave just like other animals (with the possible exception of those such as Lorenz in his classic book, *On Aggression*, published in 1967). In fact, Wilson stated, "The purpose of sociobiology is not to make crude comparisons between animal species or between animals and men, for example simply to compare warfare and animal aggression" (Wilson in Barash 1977, xiv). Many people have questioned whether such a comparison is the sole purpose of sociobiology. Yet, the animal aggression data has been useful in guiding researchers in the formulation of human experiments and, more important, in developing general theories of evolution and behavior, which then have been extended to humans. It is within this context that the animal aggression literature contributes an understanding of the biological deterministic theories of war and peace.

ANIMAL AGGRESSION AND PHYSIOLOGICAL PSYCHOLOGY

It is impossible to begin a discussion of the physiological basis of aggression without first examining what researchers have meant by the term. Although in the 1950s and 1960s several authors suggested that a simple, unitary definition of aggression was questionable, Moyer (e.g., 1968, 1974) was the first to attempt to define aggression by systematically categorizing different types. He stated that the distinctive types of aggression were not mutually exclusive and, in fact, were overlapping and could augment one another, but, he suggested, each form had different physiological bases. Further, an experimental manipulation could facilitate one type of aggression while attenuating other forms due to the stimulus-specific nature of each of the categories.

Moyer proposed eight general categories of mammalian aggression: intermale, territorial, sex-related, irritable, instrumental, maternal, fear-induced, and predatory. Although I do not subscribe to these categories, it is important to review them since many researchers in this field have been strongly influenced by Moyer's approach. The first five categories were found almost entirely among males, and maternal aggression was found only among females; fear-induced and predatory aggression did not show sex differences. Each form of aggression will be described briefly to give an idea of how it has been studied.

1. Intermale aggression. This type of aggression was tested generally by placing two previously isolated male mice in a neutral enclosure (not the home cage of either mouse). The test is problematic since group-housed animals do not show the same type of aggressive behaviors that was observed in the isolated test mice; in fact, unlike isolated mice, group-housed mice often show very little overt fighting. Further, other physiological and behavioral changes that might later promote aggression are likely to be brought about by isolation. Also, many researchers have questioned the relevance of examining isolates of a social species such as the mouse.

It is interesting to note that the behavior is not labeled and tested as "intraspecific" (that is, involving members of single species), but rather as intermale, aggression. Moyer classified it as such because he felt that both laboratory and field experiments on mammals indicated that, in virtually all cases (except hamsters and gibbons), males were highly aggressive toward conspecifics, while females showed little if any of this type of aggression. For example, in the laboratory, placing two adult, previously isolated male mice in a neutral environment will result in spontaneous fighting. Fighting does not occur, however, with two cycling or ovariectomized (noncycling) female mice or with a male and a female mouse under similar conditions.

Moyer also pointed to other fieldwork, such as that of looking at scars, wounds, and actual fighting among a variety of species, to substantiate his claims concerning intermale aggression. He stated that intraspecific aggression occurred only among males in whales, elephant seals, hippopotamuses, deer, musk ox, seals, walruses, wolves, macaques, and a variety of Australian marsupials. Interestingly, when female animals demonstrate aggressive behavior, they are often labeled as showing "sex-role reversal" (see Shaw and Darling 1985).

Moyer stated that intermale aggression is controlled largely through androgens (especially testosterone) and that it depends strongly upon pheromonal (odor) cues, which also are produced by androgens. Brain structures (e.g., the septum) are reported to be involved as well. Moyer emphasized, however, that experience is also very important, especially the experience of winning or losing a contest. (The role of experience will be examined in more detail later in this essay.)

2. Territorial aggression. This category is similar to that of intermale aggression except for a difference in the form of the test. Rather than placing two males in a neutral environment, an intruder is placed within another male's home cage territory. Fighting is assumed to be in defense of the male's territory.

3. Sex-related aggression. This type of aggressive behavior reportedly is elicited by the same stimuli that elicit sexual behavior, and almost always it is directed at the female by the male.

4. Irritable aggression. This type of aggression is again seen more in male than in female animals. It has been studied by inducing pain, often by electric shock, or by inducing frustration, often by omitting rewards in a situation in which an animal has been previously rewarded. The test animal is then provided with a conspecific. This form of aggression is not stimulus-specific (that is, the animal will attack almost anything—animate or inanimate), and it is characterized by agitated behavior and vocalization, which some researchers have termed *anger*, *rage*, and *frustration*. Irritable aggression also appears to be dependent upon androgens and various brain structures (e.g., ventromedial hypothalamus, amygdala, and septum).

5. Instrumental aggression. In this form of aggression, an animal is considered to be rewarded by an opportunity to fight with a conspecific male. Instrumental aggression overlaps with several other categories, such as intermale and territorial aggression.

6. Maternal aggression. Maternal aggression is seen when a juvenile or adult male mouse is placed in an enclosure with a parturient or lactating female mouse and her litter. This form of aggression seems to be hormonally mediated and depends upon the stimulus qualities of the young and the intruder and on the proximity of the young to the mother. This behavior is seen only in certain species and strains.

7. Fear-induced aggression. This is aggressive behavior that generally is seen when an animal is cornered by a conspecific, and it is preceded by escape attempts.

8. Predatory aggression. This type of aggression has been studied primarily by providing rats or cats with a "prey" item (mice and frogs and mice and rats, respectively) and looking for stalking, killing, and eating behaviors. Few emotional behaviors have been observed (e.g., hissing and piloerection in the cat) and, therefore, this type of behavior has often been separated from those listed above. Again, there appears

to be a neural mechanism (related to the lateral hypothalamus, amygdala, and frontal lobes) that controls the behaviors, and experience and observational learning are also involved.

Other researchers have modified or added to Moyer's categories, yet Moyer's definitions of aggression have strongly influenced the way animal aggression has been conceptualized and studied by behavioral endocrinologists and behavior geneticists. The popularity of his approach is probably due to its apparently straightforward methods; the methods all emphasize fighting and antisocial behavior, however, and, in large part, they employ testing procedures that may be context-specific and socially irrelevant.

Behavioral Endocrinology and Aggression

Hormones, particularly androgens, have long been reported to increase levels of aggression generally in a wide range of bird and mammalian species. Because of the emphasis on androgens, work in this area has focused largely on males. It is important to note, however, that female animals also produce androgens in the adrenal glands, although in smaller quantities than those produced by males. In addition to hormones related to the seasonal correlations and to the pubertal and castration effects previously mentioned in males, the roles of prenatal and prepubertal hormones in aggression have also been reported in rats and mice.

Female rats and mice that received large exogenous doses of androgen prenatally showed more aggression in adulthood in response to androgens than females that did not receive the prenatal androgens. Conversely, male rats that were castrated in utero (and, therefore, produced few androgens) were far less aggressive in response to androgens in adulthood than were normal males (see review in Leshner 1978). Further, vom Saal and Bronson reported increased levels of aggression in adulthood among female mice that were next to two males in utero as compared with female mice that were next to two females in utero (1980). They suggested that the difference relates to differences in the levels of circulating prenatal androgens. Production of prenatal androgens can be brought about in many ways. Recently, several researchers have found that prenatal stress, prompted either by heating and restraining pregnant mice (Harvey and Chevins 1985), reduced the adult levels of intermale aggression of the male offspring. Further, in the Harvey and Chevins study, normal levels of aggression were reinstated following adult testosterone treatment.

The work on the effects of prenatal hormones relates, in part, to the idea that the presence or absence of prenatal androgens leads to permanent changes in the structure of the brain, which, in turn, determines adult responses to various hormones. The theory was first conceptualized in the organization-activation hypothesis in which it was stated that the brain was organized by the presence or absence of prenatal androgen so that it would respond only in certain ways at puberty in the presence of hormones (see review in Leshner 1978). For example, it was stated that animals that were exposed to high levels of androgens at crucial times in utero would not respond in a female manner (i.e., cyclically) to estrogen/progesterone at puberty; the condition has been labeled masculinization. Similarly, animals that were not exposed to prenatal androgens would not respond to androgens at puberty—referred to as feminization.

The organization-activation theory was later modified to emphasize a prenatal sensitizing rather than organizing (e.g., vom Saal, Gandelman, and Svare, 1976; vom Saal, Svare, and Gandelman, 1976). This less deterministic theory did not state that prenatal hormones made responses to certain hormones at puberty impossible, but rather it emphasized the ease with which those responses could be seen. For example, male-like responses could be seen in adult females that did not have prenatal androgens but only with rather massive dosages of androgens over a prolonged period in adulthood.

Although the data on prenatal, pubertal, seasonal, and castration effect on aggression suggest a straightforward relationship between androgens and aggression (at least for rodents), there are many studies that do not support such a relationship. Levels of testosterone have not been correlated with intensity or amount of attack. For example, giving a castrated male rat moderate levels of testosterone will increase aggression, but low or high levels of testosterone will not (Leshner 1978).

The relationship between androgens and aggression is further complicated due to species differences with respect to the hormone that mediates the behaviors. For rats, but not for guinea pigs and rhesus monkeys, part of testosterone's effect on male behavior is a result of the conversion (or aromatization) of testosterone to estrogen, with the resulting estrogen being the active hormone.

Dominance also affects hormones. Once a mouse is at a low rank, for example, increasing the level of testosterone will not raise that animal's status. In relation to this finding, van de Poll et al. found that male rats that lost aggressive contests and were treated with testosterone were *less* aggressive than male losers that were not treated with testosterone (1982). For female rats in the study, testosterone-treated losers were more aggressive than untreated losers. Female winners treated with testosterone were less aggressive than winners not treated with testosterone.

It is also important to question the generalizability of exogenous hormones to that of endogenous hormones. In his 1983 summary of research on testosterone and strain differences in aggression, Simon said, "Little evidence has thus far been obtained to suggest that dominance is correlated with endogeneous testosterone" (21). Finally, in a recent paper, Brain and Kamis stated that different forms of aggression in mice have different endocrine bases. He refuted the assumptions that aggression is androgen-dependent and a uniquely male behavior and demonstrates that there is no simple relationship between fighting and hormones (1985).

Behavior Genetics and Aggression

A great deal of work has been done on aggression and behavior genetics in specific strains of mice. Many researchers have bred selectively for aggressive strains of male mice by using males who have demonstrated high levels of fighting in aggression tests. Although the relationship between genetics and aggression has been assumed, when it is explored empirically, the results are mixed. DeFries and McClearn found, for example, that dominant male mice sired 90 percent of offspring (1970) but Horn found that the percentage depended upon the strain of mice that were used in the experiment (1974).

In their review of aggression and behavior genetics research in mice, Hewitt and Broadhurst concluded that there is no evidence demonstrating that high levels of aggression are adaptive:

> Although our coverage has not been exhaustive, it is sufficient for us to determine that for aggressive behavior, even when we restrict consideration to same-sex (male-male) encounters in one species, the genetic architecture depends on the particular genotypes, the test procedure, type of opponent, and so on. Thus as far as the evidence from genetic architecture is concerned, the data so preclude any firm generalization about the adaptiveness of high or low levels of aggression. (1983, 59, 62)

Depending upon the strain and method of measurement, the authors reported high or low heritability of aggression and selection for low, moderate, or high levels of aggression.

Using behavior genetics techniques, Hahn examined three types of aggressive behavior in mice (1983). He found that for male-male aggression *moderate*, not high, levels of aggression were advantageous. For food getting, behavior helpful in getting food and moderate levels of aggression were advantageous. Amount of aggression was found to be totally unrelated to pup survival in the nest situation. The data again demonstrates that aggression is not a simple, unitary process with a genetic base.

This brief review of the study of aggressive processes shows that the relationship between aggression, hormones, and genes is not simple. While certain aspects of aggression seem to be related to hormones and genes in some species (primarily rodents), the relationship is a highly complex one that depends upon variables such as the definition of aggression used, the species and strain of the subjects, the manner of testing, and the animal's prior experience. The idea that aggression in nonhuman animals is *determined* by genes and hormones is therefore erroneous.

ANIMAL AGGRESSION AND SOCIOBIOLOGY

Sociobiologists have stated that animals "have been selected to behave aggressively under certain conditions and nonaggressively under others, depending on the consequences of such aggressiveness or nonaggressiveness for our evolutionary success. In short, aggressive behavior that is adaptive under one condition may be maladaptive under others; aggressive behavior that is adaptive for one individual (say, an adult male) may be maladaptive for another (say, a juvenile female)" (Barash 1986, 158).

Before further discussing the sociobiological view of animal aggression, it is necessary to briefly define the theory. Sociobiology is the study of the evolution of social behavior and is based on the Darwinian principles of variation and natural selection. Behavior is assumed to have a genetic basis (although there are few data to support this proposition), and individual animals are assumed to reproduce at differential rates due to differences in physiology and/or behavior. For example, individuals who carry a gene for a behavior that allows them to escape predation more effectively will be more likely to survive and will reproduce at a higher rate than animals that do not have such a gene. Behaviors that increase reproductive rates or maximize fitness are said to be selected for.

To this notion of natural selection was added Darwin's concept of sexual selection—that females and males of a species could have evolved under differing selection pressures for physical and behavioral traits due to their different reproductive roles (1871). Since sperm were assumed (again with no hard data) to be less costly than eggs (especially in the case of mammals), the best reproductive strategy for males, according to sociobiologists, should be to try to inseminate as many females as possible. A female, on the other hand, was assumed to invest far more in her eggs and young and therefore should be more choosy and try to maximize the resources that would be provided by her mate (territory, food, predation defense, and genetics). Sociobiologists believe that sexual selection accounts for most, if not all, sex differences in behavior.

Darwin was the first to apply the idea of sexual selection to aggression. In his discussion of mammals, he stated that, if males but not females of a species had "weapons" (e.g., horns or larger size), then these characteristics must be for intramale competition and "have been gained, partly through sexual selection, owing to a large series of victories by the stronger and more courageous males over the weaker" (1871, 831).

Parker states that, because aggression is a dangerous strategy, it would be less likely to appear in the more choosy, cautious females; it would be selected for only in males and only if it increased the insemination rate (1978). To increase the insemination rate, aggressive competition among males for female mates would clearly have been selected for. Also, if stronger, better-fed males could compete more effectively for females, competition for food resources also would have been selected for.

Some sociobiologists have discussed selection against aggression in female animals. For example, in studying social bonds and aggressive behaviors in male and female chimpanzees, deWaal said, "females may increase offspring survival probabilities by reducing agonistic tensions and maintaining group integrity" (1984, 252). He felt that the males were less influenced by social bonds and therefore showed more aggressive behaviors than female chimpanzees. Hrdy has pointed out, however, that competition among some females, particularly among primates, may not occur in lower levels than among males (1981). It is possible that competition between females may be nonovert and more subtle than that among males. Further, male-male competition may be rather transitory and intense, whereas female-female competition may be continuous and ongoing.

Barash has said that aggression "takes place when individuals interact with one another such that one of them is induced to surrender access to some resource important to its fitness" (1982, 340). Competition will be selected for when benefits outweigh costs—that is, when the resource (territory, food, or mate) is worth fighting over and the aggression does not decrease the inclusion of one's genes in future generations either through attacking kin or taking the animal away from behaviors necessary for survival (what Barash calls "aggressive neglect") (Barash 1979). He goes on to say that differential roles in reproduction between the sexes have led to differential roles in aggression: "since females tend to specialize in parenting, especially among mammals, males are free to specialize in defense of those resources that are crucial to fitness, for example, defense of the territory and protection against predators" (1982, 349). Further, due to sexual selection, females have been selected to choose more aggressive males. If aggressive males outcompete other males and if such behavior is passed on to male offspring, by choosing an aggressive mate a female increases the likelihood that her sons will reproduce, thereby increasing her genetic fitness. Again, little or no data are presented to substantiate such claims.

In addition to making similar points to those above, Wilson revised Moyer's categories of aggression to reflect adaptation and inclusive fitness in his now classic work *Sociobiology* (1975). He too proposed eight categories of aggression: dominance, territorial, sexual, parental disciplinary (toward one's young to protect them), weaning (to drive off and disperse older offspring), predatory, antipredatory (for example, mobbing), and moralistic (intentional punishment as seen only in humans). Since only the first three categories are really germane to the current discussion, only they will be discussed further.

In dominance aggression the goal is to exclude an intruder or subordinate from desired objects—those that enhance an individual's genetic fitness, including resources and mates. The competition for mates can be direct, as in male-male competition (two peacocks or stags displaying and/or fighting), or it can be more indirect, as in the case of infanticide. Sociobiologists discuss infanticide as one male killing the offspring of another male. Infanticide is seen as improving the genetic fitness of the male, who kills the young of another male because the mother is then available for mating with the new male.

Hrdy (1979, 1981), Hrdy and Hausfater (1984), and Hayssen (1984) have discussed infanticide in lions, langurs, and other species as providing a benefit to an adult male who newly gains dominance in the group. If a female is suckling her young sired by one male, she is unavailable sexually to a new male. Further, the new male may be providing protection and resources to a young animal that does not share his genes. The adult male may therefore increase his fitness by killing the young animal, which will bring the female back into estrus (the period of sexual responsiveness and ovulation). The new male will then be able to mate with the female, impregnate her, sire his own offspring, and thereby increase his fitness.

Vom Saal and Howard have examined infanticide in male mice

(1982). Sexually inexperienced males showed the highest incidence of infanticide. Among sexually experienced males, infanticide occurred if the males had engaged in sexual behavior three weeks to three months before the test. If the sexual behavior had occurred three weeks before the test (the gestation period in mice), the male demonstrated parental behavior toward the mouse pups. The authors stated that this behavior was adaptive because pups that the male could have sired were spared and pups that were probably sired by other males were killed. The authors also found, however, that there was no recognition of the female with which the male had previously mated; therefore, the behavior may or may not increase fitness. None of the studies that explore infanticide present data demonstrating increased fitness due to the infanticide.

Territorial aggression is quite similar to dominance aggression; the desired object in this case, however, is territory. Elaborate and stereotyped behaviors or displays are used especially to repulse intruders from one's territory.

The category of sexual aggression encompasses the attacks and threats used by males to force females to mate with them, which includes the incidence of so-called rape among animals. Shields and Shields (1983) and Thornhill and Thornhill (1983) have presented an argument that sexual aggression is a reproductive strategy that can enhance a male's fitness. The approach has been criticized in detail in a recent book, *Violence against Women: A Critique of the Sociobiology of the Rape* (1985), edited by Sunday and Tobach (especially in the prologue and epilogue). The authors' major criticisms center on the definitions of rape that are used by sociobiologists and the lack of data used to substantiate the claims of increased fitness for the male who rapes.

Many other excellent general critiques of sociobiological theory have been written (for example, Montagu 1980; Lewontin, Rose, and Kamin 1984; and Kitcher 1985). The authors have pointed out that sociobiological analyses of behavior turn into "just-so stories" in that they are either untestable or the data are not presented. Little or no evidence has been provided that behavior is genetically controlled, even among nonhuman species, and little or no data have been presented that demonstrate increases in inclusive fitness. Universality of behavior, even within a species (not to speak of between species), has not been demonstrated sufficiently. The sociobiological view that aggression has been selected for among male animals is, therefore highly problematic.

SUMMARY AND CONCLUSIONS

Those who view human aggression as being biologically determined draw heavily from the data concerning the biological basis of aggression among nonhuman species. That work has centered on three basic approaches: behavioral endocrinology, behavior genetics, and sociobiology. There are numerous problems in all three of these approaches, which assume that biological factors *determine* aggression. First, there is a problem in defining aggression. Aggression is not a single motivational process that can be defined universally, even among nonhumans. It depends upon the species and strains used and the testing procedures employed. Second, the vast majority of testing has focused on domesticated mice and rats—species that may tell us little that can be generalized to other species. Third, the procedures used in the studies do not employ socially relevant testing. Social species are generally isolation-reared in the laboratory, encounters are measured by short-term bouts between individual animals as opposed to groups of animals, and the measures that are studied are fighting and tissue trauma rather than longer-term social interactions.

Researchers who assume that aggression is biologically determined believe that animals fight because they are driven by instinct (e.g. Lorenz), their hormones, or their genes. Although the research that these groups of scientists have conducted has been on nonhuman animals, the generalization is made often that, since humans are also animals, they will behave like other animals. This supposition often leads to the assumption that human aggression is an outcome of our genes and hormones and is inevitable. At best, the hormonal, genetic, and sociobiological data indicate that there is no simple relationship between biology and aggression among nonhumans. Add to this the extreme problem of generalizing from the laboratory to the natural environment and from nonhuman species to humans, and it is very clear that, while aggression may be affected by biological factors, it is not biologically determined. Aggression is neither inevitable nor uncontrollable in human or nonhuman species.

REFERENCES

Barash, D. P. 1977. *Sociobiology and Behavior*. New York: Elsevier.
______. 1979. *The whisperings within*. New York: Harper and Row.
______. 1982. *Sociobiology and behavior*, 2d ed. New York: Elsevier.
______. 1986. *The hare and the tortoise: Culture, biology and human nature*. New York: Viking.
Brain, P. F., and A. Kamis. 1985. How do hormones change "aggression"? The

example of testosterone. In *Aggression: Functions and causes*, edited by J. M. Ramirez and P. F. Brain, 84–115. Seville, Spain: Publicationes de la Universidad de Sevilla and Professors' World Peace Academy.

Darwin, C. 1871. *The descent of man and selection in relation to sex*. New York: Modern Library.

DeFries, J. C., and C. E. McClearn. 1970. Social dominance and Darwinian fitness in the laboratory mouse. *The American Naturalist* 104:408–11.

deWaal, F. B. M. 1984. Sex differences in the formation of coalitions among chimpanzees. *Ethology and Sociobiology* 5:239–55.

Hahn, M. E. 1983. Genetic "artifacts" and aggressive behavior. In *Aggressive behavior: Genetic and neural approaches*, edited by E. C. Simmel, M. E. Hahn, and J. K. Walters, 67–88. Hillsdale, N.J.: Lawrence Erlbaum.

Harvey, P. W., and P. F. D. Chevins. 1985. Crowding pregnant mice affects attack and threat behavior of male offspring. *Hormones and Behavior* 19:86–97.

Hayssen, V. D. 1984. Mammalian reproduction: Constraints on the evolution of infanticide. In *Infanticide: Comparative and evolutionary perspectives*, edited by G. Hausfater and S. B. Hrdy, 105–23. New York: Aldine.

Hewitt, J. K., and P. L. Broadhurst. 1983. Genetic architecture and the evolution of aggressive behavior. In *Aggressive Behavior: Genetic and neural approaches*, edited by E. C. Simmel, M. E. Hahn, and J. K. Walters, Hillsdale, N.J.: Lawrence Erlbaum.

Horn, J. M. 1974. Aggression as a component of relative fitness in four inbred strains of mice. *Behavior Genetics* 4:373–81.

Hrdy, S. B. 1979. Infanticide among animals: A review, classification, and examination of the implications for the reproductive strategies of females. *Ethology and Sociobiology* 1:13–40.

______. 1981. *The woman that never evolved*. Cambridge, Mass. Harvard University Press.

Hrdy, S. B., and G. Hausfater. 1984. Comparative and evolutionary perspectives on infanticide: Introduction and overview. In *Infanticide: Comparative and evolutionary perspectives*, edited by G. Hausfater and S. B. Hrdy, xiii–xxxv. New York: Aldine.

Kingsley, C., and B. Svare. 1986. Prenatal stress reduces intermale aggression in mice. *Physiology and Behavior* 36:83–86.

Kitcher, P. 1985. *Vaulting ambition*. Cambridge, Mass.: MIT Press.

Leshner, A. I. 1978. *An introduction to behavioral endocrinology*. New York: Oxford University Press.

Lewontin, R. C., S. Rose, and L. J. Kamin. 1984. *Not in our genes: Biology, ideology and human nature*. New York: Pantheon.

Lorenz, K. 1967. *On aggression*. New York: Bantam.

Montagu, A., ed. 1980. *Sociobiology examined*. New York: Oxford University Press.

Moyer, K. E. 1968. Kinds of aggression and their physiological basis. *Communications in Behavioral Biology* 2:65–87.

______. 1974. Sex differences in aggression. In *Sex differences in behavior*, edited by R. C. Friedman, R. M. Richart, and R. L. VandeWiele, 335–72. New York: John Wiley and Sons.

Parker, G. A. 1978. Assessment strategy and the evolution of fighting behaviour. In *Readings in sociobiology*, edited by T. H. Clutton-Brock and P. H. Harvey, 271–92. New York: W. H. Freeman.

Shaw, E., and J. Darling. 1985. *Female strategies*. New York: Walker.

Shields, W. M., and L. M. Shields. 1983. Forcible rape: An evolutionary perspective. *Ethology and Sociobiology* 4:115–36.

Simon, N. G. 1983. New strategies for aggression research. In *Aggressive behavior: Genetic and neural approaches*, edited by E. C. Simmel, M. E. Hahn, and J. K. Walters, Hillsdale, N.J.: Lawrence Erlbaum.

Sunday, S. R., and E. Tobach, eds. 1985. *Violence against women: A critique of the sociobiology of rape*. New York: Gordian Press.

Thornhill, R., and N. W. Thornhill. 1983. Human rape: An evolutionary analysis. *Ethology and Sociobiology* 4:137–73.

van de Poll, N. E., J. Smeets, H. G. van Oyen, and S. M. van der Zwan. 1982. Behavioral consequences of agonistic experience in rats: Sex differences of the effects of testosterone. *Journal of Comparative and Physiological Psychology* 96:893–903.

vom Saal, F. S., and F. H. Bronson. 1980. Sexual characteristics of adult female mice are correlated with their blood testosterone levels during prenatal development. *Science* 208:597–99.

vom Saal, F. S., R. Gandelman, and B. Svare. 1976. Aggression in male and female mice: Evidence for changed neural sensitivity to neonatal but not adult androgen. *Physiology and Behavior* 17:53–57.

vom Saal, F. S., and L. S. Howard. 1982. The regulation of infanticide and parental behavior: Implications for reproductive success in male mice. *Science* 215:1270–72.

vom Saal, F. S., B. Svare, and R. Gandelman. 1976. Time of neonatal androgen exposure influences length of testosterone treatment required to induce aggression in adult male and female mice. *Behavioral Biology* 17:391–97.

Wilson, E. O. 1975. *Sociobiology: The new synthesis*. Cambridge, Mass.: Belknap Press of Harvard University Press.

♦ SUSAN OYAMA ♦

Essentialism, Women, and War: Protesting Too Much, Protesting Too Little

Recently some biological theorists and feminists have converged on "essentialist" accounts of war that are strangely similar in certain ways. At first glance it seems to be an unlikely development, given the frequency with which we have seen antifeminists and feminists line up on opposite sides of the nature-nurture rift. At second glance, though, perhaps the association is not so surprising after all. We seem to be in the midst of a pendulum swing "back to nature" and away from environmentalism. This movement, in turn, is probably part of a more general trend in this country toward conservatism and a certain brand of romanticism, though the issue is a good deal more complex than one might think. Apart from the current emphasis on so-called traditional values, though, the convergence on "biological" views reflects some very common and pervasive beliefs about genes and environment, biology and learning, that are as evident in environmentalist approaches as they are in biological ones.

By essentialist, I mean an assumption that human beings have an underlying universal nature, one that is more fundamental than any variations that may exist among us and that is in some sense always present—perhaps as "genetic propensity"—even if it is not discernible. People frequently define this preexisting nature in biological terms, and they believe it will tend to express itself even though it might be somewhat modified by learning and thus might be partially obscured by a sort of cultural veneer. (For a good discussion of this theme in feminism, see Alison Jaggar [1983, chap. 5]. Janet Sayers [1982, 148] criticizes essentialist feminism, and Anne Fausto-Sterling [1985, 195] refers to human sociobiology as a "theory of essences"; Ruth Bleier [1984, chap. 1] criticizes both. Of these authors, Jaggar is perhaps most successful in transcending the biology-culture opposition, but all are aware of the mischief it has caused for scientists and nonscientists alike.)

When I say that environmentalist and biological approaches share many assumptions about nature and nurture, I mean that they have often argued about which and how many traits were genetic and which were learned, but, in doing so, they have accepted the premise that genes and learning were properly treated as alternative explanations for human characteristics and actions. They also tended to agree that the possibility of change was somehow illuminated by their disputes.

In a metaphor that is revealing in more than one way, sociobiologist David Barash compares the relationship of nature and nurture to two people wrestling. As they tumble about, their limbs entwine so that it is hard to tell which is which. However entangled they may become, the combatants do not merge; they are separate persons in competition, and our imperfect powers of observation do not change that fact (1981, 12).

Though they routinely declare that the nature-nurture dichotomy is meaningless and that the effects of biology and culture cannot be clearly distinguished, scholars of all stripe (including not only sociobiologists like Barash but many of sociobiology's critics as well—see discussion and references in Oyama 1981, 1982) continue to treat them as separate sources of living form and behavior: some things are (mostly) programmed by our genes, others are (mostly) programmed by our environments. We will return to this conceptual problem later.

Let's look first, however, at some examples of essentialist accounts of women and war. The first several examples come from scholars who have offered us their biological views of human behavior and society, while the last two come from a recent collection of feminist writings.

THE ARGUMENT

Lionel Tiger and Robin Fox say that war "is not a human action but a male action; war is not a human problem but a male problem." If nuclear weapons could be curbed for a year and women could be put into "all the menial and mighty military posts in the world," these authors declare, there would be no war. They immediately concede that this proposition is but a fantasy, and a totally unrealistic one at that, because the human "biogrammar" (a term they use more or less the way others use "genetic program") ensures that such a thing could never happen. Men, they say, have evolved as hunters who band into groups and turn their aggressiveness out against common enemies or prey (1971, 212–13). Political structures in modern societies are formed on this primeval hunting model, and, naturally, men dominate these structures as well. Tiger speculates that women, who do not bond and cooperate as effectively as men, could be given positions in government by special mandate. He feels, however, that it might be quite unwise to expect that even this effort could effect much change (1970, 270–72); presumably, attempts to subvert biologically natural tendencies are not likely to succeed.

Barash accepts the idea that males bond and exclude women from political power (1981, 187–89), though he emphasizes the grounding of male aggression in the competition for reproductive opportunities and reproductively relevant resources (174, for example). He argues that women are only allowed political power if they are in some sense "desexed," by age and/or physical unattractiveness. Otherwise, men refuse to recognize a woman's authority, even when she manages to gain admittance to the male "club" (189–90).

Another writer on biological topics, Melvin Konner, also suggests that, because they are less aggressive than men, women should be placed in authority in order to "buffer" or "dampen" violent conflict between nations (1982, xviii, 126, 420). He seems to reject hunting bands as the evolutionary explanation for human aggression, though he does cite Tiger's work and allows that "something happens when men get together in groups; it is not well understood, but it is natural, and it is altogether not very nice" (203–206). Like Barash, Konner is more impressed with the notion that male aggression is explained by competition for the reproductive resources provided by females: eggs and parental care (chap. 12).

The suggestion that women might be more peaceful than men in positions of power is thus immediately and quite effectively undone by the theorists' other assumptions about natural differences between women and men, and the social consequences of these differences. It is a rather neat irony—that the qualities that might save the world are kept out of the public sphere by the very biological order that produces them. (Konner does not say women must always be excluded from power, though he says that it is pointless to use violent female rulers of the past as models for the future because they "have invariably been embedded in and bound by an almost totally masculine power structure, and have gotten where they were by being unrepresentative of their gender" [1982, 126]. He does not say how to implement his suggestion that "average" women be allowed to control the world's arsenals.)

The argument that women are inevitably excluded from political life also appears, of course, in past and present antifeminist writings on the necessity of patriarchy (see Sayers 1982, for discussion; and Goldberg 1973, for a relatively recent example). Partly because biological arguments have often been associated with reactionary politics, it is now common for theorists to declare their liberal values, deny that biological treatments are necessarily either deterministic or conservative, and emphasize that biological explanation is not the same as moral approval. Then the theorists typically call on us to know our natures in order to transcend them. Barash, for example, asks whether we can use our understanding to overrule the biological "whisperings from within" (1982, 198). At the same time, they often warn against trying to challenge the boundaries and constraints our genes set for us. Charles Lumsden and E. O. Wilson warn that trying to escape these constraints risks the "very essence of humanness." They advise us to learn what the limits are and to set our goals within them (1981, 359–60), while Barash declares that denying natural sex differences is "likely to generate discontent" (1981, 116; for a critique of the language of constraints and limits, see Oyama 1985).

The relationship between politics and science is a complex one, and, though I think it can be argued that at any particular time some scientific approaches tend to be associated with and/or imply certain attitudes toward the moral and political worlds, there is no direct link between reactionary values and an interest in, for example, sociobiological analysis. Biologists become quite as annoyed at having their politics misrepresented as anyone else, and to assume that someone who emphasizes biological bases of human behavior is automatically a "crypto-nazi" is to engage in just the sort of reductionist thinking I am criticizing in this paper. A crucial link between one's scientific and political views is one's conception of will and possibility, and this notion is rarely made explicit. Just because the relationships are so complicated, however, it becomes very important to make assumptions explicit whenever possible, for it is these hidden assumptions that structure the arguments and invite the conclusions.

Examples of feminist essentialism are found in Pam McAllister's collection, *Reweaving the Web of Life* (1982). In "The Prevalence of the Natural Law within Women," Connie Salamone describes women's roles in protecting both the species' young and the natural law that governs the world. This role, if it is not subverted by male values, endows females with a special affinity to other animals and tends to give rise to concern over animal rights and vegetarianism. Salamone contrasts the female "aesthetic of untampered biological law" with "the artificial aesthetic of male science" (1982, 365–66). In "Patriarchy: A State of War," Barbara Zanotti invokes Mary Daly's concept of women's *biophilia* (love for life) and describes the history of patriarchy as the history of war. She asserts that, in making war, patriarchy attacks not the opposing military force but women, who represent life. Soldiers, she suggests, are encouraged to identify military aggression with sexual aggression, so that "the language of war is the language of gynocide" (1982, 17).

Here I have addressed some versions of the argument: that women are inherently less aggressive than men, war is caused by male aggression, and women are thus somehow more capable than men of bringing about peace—or, at least, if they were in power, women would be less destructive than men. It is a skeletal argument, of course, abstracted from very different sorts of writings.

The transition from individual to international conflict is not necessarily direct; for Tiger and Fox (1971), Barash (1981), and Zanotti (1982), for example, war is specifically the aggression of male-bonded men in groups, and Tiger (1970, 219) distinguishes between individual and group violence. Furthermore, the connection between aggression and peacemaking is not clear. Especially in the work of the male scientists cited above, it is *lack* of aggressiveness, rather than any positive quality, that is emphasized. Even in a world in which women are traditionally defined by their deficits, I'm not sure that peacemaking and peacekeeping should be seen as merely passive (to invoke another loaded dichotomy) results of low levels of aggression. Recent radical feminists are more likely to point out positive female qualities of nurturance, sensitivity to connections, and peacefulness. (I follow Jaggar 1983, chap. 5; and Sayers 1982, in using the term *radical* here. For the purposes of this essay, it entails a tendency to speak of essential feminine qualities in a positive, even celebratory way, rather than insisting on women's basic similarities to men.)

LUMPING: HOW TO IGNORE IMPORTANT DISTINCTIONS

The arguments all require that aggression be somehow unitary. They depend on, and encourage, certain kinds of illegitimate "lumping." One sort of lumping is definitional; that is, all sorts of behavior, feelings, intentions, and effects of actions are grouped together as aggressive. Tiger's definition, for example, is so broad that it embraces all "effective action which is part of a process of mastery of the environment"; violence is one outcome of such aggressive activity, but not the only one (1970, 203). That women do not bond in aggressive groups implies, then, that they are less capable of "effective action" in the service of "mastery of the environment"—a sweeping statement indeed, and one that bodes ill for any political action on the part of women.

Another approach that allows us to treat aggression as a uniform quantity is cross-species lumping. Very different phenomena often are equated so that, for example, mounting or fighting in rodents, territoriality in fish or birds, and hunting, murder, or political competition in humans are all "aggressive" behaviors. Then there is the lumping of levels of analysis. The activity of nations and institutions is reductively collapsed to the level of individuals or even of hormones or genes. Finally, there is developmental lumping. Activity levels in newborn babies, rough-and-tumble play in young children, fighting or delinquency in teenagers, and decisions of national leaders in wars are viewed as somehow developmentally continuous, or "the same." Sex differences in these behaviors are then seen as manifestations of basic sex differences in aggressiveness (see Money and Erhardt 1972 for a flawed but highly influential treatment of some sex differences; see also critiques by Bleier 1984; Fausto-Sterling 1985; and this and other volumes in the Genes and Gender series; also see Klama 1988 for treatment of more general issues in aggression studies).

Much of this lumping depends on very common modern-day versions of preformationism and essentialism. Today we think of preformationism as an archaic relic of outmoded thought, and we snicker about the absurd idea that there could be little people curled up in sperm or egg cells. But replacing curled-up people with curled-up blueprints or programs for people is not so different. That is, whether we speak of aggression in the genes or coded instructions for aggression in genes, we haven't made much conceptual progress. What is central to preformationist thought is not the literal presence of fully formed creatures in germ cells but, rather, a way of thinking about development—development as revelation of preformed nature or essence, as expression of preexisting program or plan, rather than as contingent series of constructive interactions, transformations, and emergences. It is a view that makes real development irrelevant, since the basic "information," or form, is there from the beginning, a legacy from our evolutionary ancestors (see Oyama 1985, for fuller treatment of these issues).

Nor is the basic reasoning much changed by the less-deterministic-sounding language of biological predispositions, propensities, or limits to flexibility; the assumptions underlying these apparently more moderate formulations are not substantially different from the more dichotomous ones that people ridicule these days. Similarly, saying that, of course, nature combines with, or interacts with, nurture shows continued reliance on a biological nature defined by genes before development begins and moderated or deflected by an external, environmental nurture. Even though no one claims our natures to be absolutely uniform and immutable, the somewhat softer language of genetic predispositions and tendencies shares the logical weaknesses of strict determinism, even if it seems to give us more possibilities for change. One problem is that, in this more moderate sort of account, the genes still define the boundaries within which action is possible, and they still constitute the ultimate source of control. (Barash likens us to horses being ridden by genetic riders who give us considerable freedom but who remain in firm command [1981, 200].) A puzzle for those who hold this view is how to conceptualize a "we" that is pitted against our genes in a struggle for control over "our" behavior. Another puzzle, for feminists who embrace the argument for inherent male aggression and dominance, is how to mobilize for change in a world populated by inherently aggressive and dominant males.

WHAT IS THE POINT OF THIS CRITIQUE?

It is important to be clear about what I am *not* doing when I critize the nature-nurture opposition or the lumping of different definitions, species, levels of analysis, and developmental phenomena that often accompanies it. I am not saying that aggression, however it is defined, is unimportant. I am not denying that nations are composed of individuals and that individuals are composed of cells, chemicals, and so on. I am not denying that understanding these parts might help us to understand the wholes of which our world is composed. I am not, therefore, rejecting research on individuals, hormones, neurons, and genes, including those of other species. I am not making an environmentalist argument—that biology is irrelevant, that genes don't count, and that everything about our behaviors can be changed (these three are not the same argument, and one of our problems is that we tend to lump them at the same time that we lump biological arguments). I am not even denying certain constancies or similarities among individuals within and across societies, though often we are rather cavalier with our methods of demonstrating these likenesses.

What I *am* saying is that analysis should be conducted in the interests of the eventual synthesis of a complex, multilevelled reality (just as temporary lumping—of diverse essentialist treatments of aggression, for example—can serve the elaboration of a more complex argument). The levels I have in mind here are not like onion skins that can be stripped away to reveal a more basic reality. After all, when you take away enough of an onion's layers, there's nothing left to reveal. Rather, they are levels of analysis whose interrelationships must never be assumed, but discovered. We will never understand the role of genes and hormones in individual lives or of individuals in society unless we move beyond traditional oppositions. We will never gain insight into the possibilities of different developmental pathways if we assume them to be fixed on the basis of an inappropriate argument. This is the point at which the environmental and biological determinists, as well as the more moderate "in-betweenists" are unwitting allies: they usually agree on what it would *mean* for something to be biological or cultural even as they argue about relative contributions of genes and learning.

If we want to use scientific analysis to answer questions, we must know what questions we are asking, or we'll never know what evidence could help us answer them. And if we want to fight the good fight, we must know what the enemy is, or we will waste precious time and energy that may not be ours to waste. (Note; I have said *what* the enemy is, not *who*, because in this case I am concerned with ways of thinking, not people, that make our task harder.)

QUESTIONS, CONCERNS, AND ANSWERS

We reveal a great deal of confusion when we ask if something is biological. We might be asking about the chemical processes associated with some behavior, for instance—this is a matter of the level of analysis, and such questions can be asked about any behavior, learned or unlearned, common or uncommon, fixed or labile. We might be asking about development: Does a given behavior, for example, seem to be learned? Is it present at birth? (These questions are not the same thing, since learning can be prenatal.) We might be asking about evolutionary history, which in turn, resolves into several kinds of questions: Is the behavior present in phylogenetic relatives? When did it appear in our own evolutionary line, and why? We might be asking what role, if any, a character now plays in enhancing survival and reproduction. We might be asking whether variation in a character is heritable in a given population (whether differences in the character are correlated with genetic differences). This last question has to do with population genetics, which are useful if one wants to know about the possibility of artificial or natural selection in the population.

These are very different questions for which different evidence is relevant, and they do not exhaust the catalog of biological queries.

None has any automatic bearing on any other, and lumping them together as genetic or biological simply creates confusion and faulty inferences. Often, however, a person asking whether some trait is biological is not interested in these particular questions at all but has something else in mind. She or he is concerned about the inevitability of a trait, or its unchangeability in the individuals evincing it, or its goodness, justifiability, or naturalness, or perhaps the consequences of trying to change or prevent it (will it come bursting out as soon as we drop our guard? Will intervention do more harm than good?). Scientists frequently share these concerns and the confusions that link them to biology.

None of the scientific questions listed earlier—about evolution, developmental timing, or process or level of analysis—is relevant to these underlying concerns. Our misguided but deeply embedded beliefs about genes and biology, however, cloud the issue. *Genetic* and *biological*, in fact, are often effective synonyms for *inevitability, unchangeability,* and *normality.* The common concept of genetic control and guidance of development implies that fate, or at least the range of potential fates, is set before birth. But persons must develop, and development is the result of a whole system; there is no clear way to see the role of genes as more basic, formative, directive, controlling, or limiting than other aspects of system, and the role of any particular factor depends on its interrelations among the others.

Concerns about inevitability are really about *possible* developmental pathways, not about past or present ones. (Even wondering whether a present state of affairs is immutable implies wanting to know *what would happen if* . . . , and wondering whether it was inevitable implies wanting to know *what would have happened if* . . . .) When Barash speculates that male parenting in humans, as in other mammals, is "not nearly as innate as modern sexual egalitarians" think, he seems to be saying something about the probability of reaching certain personal and political goals, and he certainly seems to believe that "innateness" (a concept he never defines satisfactorily) has something to do with the difficulties that he thinks "sexual egalitarians" will encounter (1981, 88).

Because any pathway is the function of an entire developmental system, which includes much more than genes, its qualities are not predictable from genes alone. We would have to understand development extremely well to know whether the necessary conditions for constancy or change in patterns of aggressive behavior would be present in any particular alternative world. To say anything intelligible about possible relationships among nations, we would need to know a great deal about issues that are not in any simple way related to particular sorts of individual aggressiveness.

Inevitability is not predictable from observations at the morphological or biochemical levels of analysis. It is not predictable from the role of learning in the development of a behavior or from its time of appearance. It is not predictable from phylogenetic history, a pattern of heritability in some population, prevalence in certain environments, or even universality. That is, none of these traditional scientific biological questions is relevant to the concerns that most often motivate the questions. To ask biology to address concerns about desirability, furthermore, is to ask science to do our moral work for us.

We must decide what kind of world we want, and why. We won't necessarily succeed in bringing it about, but we shouldn't be deterred prematurely from trying because of biological evidence of whatever variety, either because we believe the biological, in any of its senses, is fixed or because we believe it is dangerous to tamper with what we think of as natural. Similarly, we shouldn't be complacent about natural features we might value (virtues that are thought basically feminine in this world won't necessarily persist in the one that's coming). There is a tendency to view the biological as static, but it is, in fact, historical at all levels. When I say "history," I am referring to contingency, interaction, possibility, and change. (The habit of asking whether some feature of our world is the result of biology or history is thus deeply mistaken). When we ask about biology, though, our concerns tend to be mythological, not historical. Here I do not mean myth as wrong, or "bad," science (though it might be) but as a way of thinking that reveals ultimate truth, eternal necessity, and legitimacy.

Lionel Tiger, chronicler of male bonding and aggression, refers to *Lord of the Flies*, the widely read story of a group of English schoolboys. Marooned on an island, the lads rapidly degenerate into a horde of savage little creatures. Apparently, the author of the book, William Golding, has said he wanted to construct a myth, a tale that would give the key to the whole of life and experience (cited in Tiger 1970, 207). The feminist Zanotti, too, accepts the centrality of male bonding to individual and social life in her claim that, in making war, men are eternally attacking and destroying women (1982, 17). Both theorists invoke unchanging essence to explain gender, relations between men and women, and, thus, the world. But it is a static world in which ancient tragedies are played out again and again according to primal necessity, not a historical world in which necessity and nature arise by process and then give way to other necessities and natures. Nature, then, should not be seen as one term in the traditional nature-nurture, genes-environment, biology-culture pair. It is not a *cause* of development but rather an emerging *product* of development.

PLAYING THE GAME

It should be clear by now how I feel about several common strategies for dealing with biological arguments. When someone says, "It's biological," we reply "No, it's not, it's cultural," when instead we should be asking why the cultural and the biological are treated as alternatives in the first place and just what we (and they) really mean by either explanation. I call this the Protesting Too Much Syndrome, because we are often afraid the trait in question *is* biological in one or more of the mistaken senses described above. Or someone says that we are innately inferior, and we counter, "No, we're not, *you* are," rather than rejecting the assumption of essential nature that allows *any* pronouncements of this sort. This second strategy I have dubbed the Protesting Too Little Syndrome. It entails agreeing that differences *are* biological, but reversing the evaluative polarity. Male nature is bad; female nature is good. While it offers the momentary satisfaction of turning the tables, it is based on all the mistaken ideas about nature and nurture that, I think, get us all into so much difficulty—too great a price to pay for Mother Nature's favor. The solution is not to protest precisely the correct amount nor to find the degree of biological determination that is "just right," like some Goldilocks trying to find comfort in a house that is not her own. Rather, it is to protest a whole lot *about the very rules of the discourse*. We mustn't allow the argument to be defined for us. Instead, we must be reflective enough to rethink it.

I am not saying that we ought to throw out everything and start from the ground up. We couldn't do it if we wanted to. But when there are ample grounds for doubting the validity of a conceptual framework or a set of issues, as is the case with the nature-nurture complex, we do ourselves no favor by blindly accepting the terms of the game. Some of our gravest problems come, after all, from letting others set terms for us. The burden of clarification certainly does not rest entirely with women, but, if we shirk our part, how can we do justice to the struggle?

Instead of pitting one mythical account against another, instead of searching for a morally or emotionally resonant evolutionary past to explain the present and then projecting it into the future, we must focus on real historical processes whose courses are not foreseeable on the basis of any account of nature as manifested in hunter-gatherers, baboons or chimps, hormones, brain centers, or DNA strands. I speak here of individual developmental history as well as historical change on the societal level, for it is within these processes that nature and possibility are defined.

ARE WOMEN LESS AGGRESSIVE AND, HENCE, LESS WARLIKE?

I could say much about aggression and about women and maybe even a little about war, but, in this essay I haven't said much about any of them. Perhaps the reason is that the essentialist theories I have been discussing don't say much about these topics either. Instead, I have focused on the ways we think about these topics. War is about politics, diplomacy, economics, and historical continuity and change in relations among people, not about brain centers, testosterone, or rough-and-tumble play. It is like a fight between individuals only by analogy, just as certain encounters between groups of ants is war only by analogy. Perhaps it is significant that when sociobiologist David Barash coauthors a book on preventing nuclear war (Barash and Lipton 1982), it contains nothing about different capacities and contributions of males and females but instead gives lists of very pragmatic suggestions for effective action. Obviously, I would never claim that women have no role in national and international politics, but neither can I make sense of the notion that we ought to be somehow inserted into public life because of some mythic direct line to life, peace, and love. Men are not a plague, and women are not a cure.

REFERENCES

Barash, D. P. 1981. *The whisperings within*. New York: Harper and Row.

Barash, D. P., and J. E. Lipton. 1982. *Stop nuclear war!* New York: Grove Press.

Bleier, R. 1984. *Science and gender: A critique of biology and its theories on women*. New York: Pergamon Press.

Fausto-Sterling, A. 1985. *Myths of gender: Biological theories about women and men*. New York: Basic Books.

Goldberg, S. 1973. *The inevitability of patriarchy*. New York: Morrow.

Jaggar, A. M. 1983. *Feminist politics and human nature*. Reprint. Totowa, N.J.: Rowman and Allanheld.

Klama, J. 1988. *Aggression: The myth of the beast within*. New York: Wiley.

Konner, M. 1982. *The tangled wing: Biological constraints on the human spirit*. New York: Holt, Rinehart and Winston.

Lumsden, C. J., and E. O. Wilson. 1981. *Genes, mind, and culture*. Cambridge, Mass.: Harvard University Press.

McAllister, P., ed. 1982. *Reweaving the web of life*. Philadelphia: New Society Publishers.

Money, J., and A. A. Ehrhardt. 1972. *Man and woman, boy and girl*. Baltimore: Johns Hopkins University Press.

Oyama, S. 1981. What does the phenocopy copy? *Psychological Reports* 48:571–81.

———. 1982. A reformulation of the concept of maturation. In *Perspectives in ethology*, vol. 5, edited by P. P. G. Bateson and P. H. Klopfer, 101–31. New York: Plenum.

———. 1985. *The ontogeny of information: Developmental systems and evolution*. Cambridge: Cambridge University Press.

Salamone, C. 1982. The prevalence of the natural law within women: Women and animal rights. In *Reweaving the web of life: Feminism and nonviolence*, edited by P. McAllister, 364–75. Philadelphia: New Society Publishers.

Sayers, J. 1982. *Biological politics*. London: Tavistock.

Tiger, L. 1970. *Men in groups*. New York: Vintage Books.

Tiger, L., and R. Fox. 1971. *The imperial animal*. New York: Holt, Rinehart and Winston.

Zanotti, B. 1982. Patriarchy: A state of war. In *Reweaving the web of life*, edited by P. McAllister, 16–19. Philadelphia: New Society Publishers.

♦ SUSAN G. GORDON ♦

Discussion

Man the aggressor and woman the peacemaker. How many times have we heard this theme? Men are "naturally" aggressive and warlike, and thus wars are inevitable; women are "naturally" peaceful and therefore antiwar. The theme is constantly repeated, and we are taught it consistently from our earliest reckoning (yes, even before our understanding) on through secondary school, higher education, and into postgraduate studies and professional schools—to the boys: you must be a man (strong and aggressive); and to the girls: be feminine (sweet and passive). Although much of this folklore is handed down through parents, grandparents, friends, teachers, and society in general, it is also given a pseudoscientific basis by research designed to prove preconceived societal ideas about human behavior. Even some of our sisters have picked up the dogma of genetic and hormonal determinism of human social behavior and advocate world leadership by naturally peaceful women so that there will be no more wars.

Most biologic and social science research that is done in North America or that is widely referred to in our literature is Eurocentric, reflecting Western white male values, theories, and belief systems. By this means we have effectively ignored the contributions of people of color from Asia, the Middle East, Africa, Latin America, and Pacific Island societies, whose scientific, scholarly, and creative contributions in some instances predate Western civilization by hundreds of years, and whose traditions may or may not be aggressive (as we define the term). I am reminded particularly of Eleanor Leacock's study of the Montagnais-Naskapi of Canada (most likely originating in Asia) who had an egalitarian society in precolonial days. Although there was division of labor according to gender, all tasks were of equal importance and there was often role reversal (men assuming child care duties, for example). The men of the tribe were greatly chided by the French monks (among the earliest European settlers of Canada) for not being aggressive enough and "allowing" the women of the tribe to be considered essentially equal. (For more information on these studies, see the selected bibliography in this volume.)

Dominant community biases inform biologic and social science knowledge production and prevail in the very designing of what research is to be done, what theories and postulates are formed and are to be tested, what tools are needed, what data are to be collected, and how the results are to be evaluated and determined. Most of the research and theorizing on aggression is done on lower animals and some primates and is simply extrapolated to modern day human beings. If ethologist Konrad Lorenz could obtain such recognition in hypothesizing the origin of human social behavior, why not prominent sociobiologists Moyer, Barash, and Wilson (who denies it) among others?

The three papers in this section present and discuss many of these points. In discussing the research work of the physiological psychologists, Suzanne Sunday neatly points out pitfalls in the formulation of their theories: their inconsistent definitions of aggression; the problematic laboratory experimentation in their research designs and observations; the inconsistent findings and conclusions of various researchers; and the dangers inherent in trying to extrapolate behavioral data from observation of lower animals, mainly rodents, and apply it to human beings. Sunday analyzes the sociobiological theories regarding aggression in male animals, which state that aggression enables fitness and is genetically transmitted up the evolutionary scale. Her analysis reveals the pseudoscientific nature of the work of Wilson, Barash, and other sociobiologists, which is primarily based on assumption and perpetuates the onerous burden of genetic determinism.

Betty Rosoff reviews and analyzes data gathered from studies on the effects of male and/or masculinizing hormones such as androgens, including testosterone, and certain synthetic progestins in relation to humans. She also points out biased research designs and inconsistent findings, and reminds us of several studies that came to erroneous conclusions as a result of biases and faulty research (for example, studies on XXY individuals and criminal behavior).

Susan Oyama's paper is a fascinating discussion of the theories and "evidence" for "men as warriors and women as peacemakers" and shows us the convergence of some biological and feminist theories concerning war and peace. Oyama calls this "essentialism," that is, belief in an "underlying universal nature"—another euphemism for genetic determinism. Her critique of the essentialist position, particularly that espoused by feminists, is very important for all feminists who might not see the snares within it. She makes it quite clear that to reduce war and peace to an essentialist position is, in effect, a reductio ad absurdum. Her recognition that the whole problem is multileveled and complex is timely and useful. She says that our immediate concern ought to be asking the right questions and doing the necessary work to answer them. Instead, we get bogged down in arguments about whether women should have power because their basic natures are peaceful.

All three of these papers recognize that human social behavior is complex, multifaceted, multicausal, reciprocal, and flexible. Sociobiologists' facile assumption of a mechanical and stereotypical duplication in human beings of lower animals' much simpler patterns of behavioral organization is not only insulting but has no basis in scientific fact (after all human beings have fifty billion brain neurons and chimpanzees only five billion). The conceptual and transformational capacities of the human brain supersede the influence of any other tendency as the basis of complex behaviors such as intelligence, gregariousness, cooperation, competition, directed aggression, social conflict, and war.

Human consciousness is determined by social interaction and results in motives, will, determination, choice of objectives, love, and hate; consciousness informs and shapes all complex social behaviors in human beings. These capacities not only give humans greater degrees of freedom than lower animals experience; they also place on human kind greater responsibility for its behavior.

♦ II ♦

Assigning Gender to Peace and War: Social Processes

♦ REGINA E. WILLIAMS ♦

Blindsight

The eyes you stretch to the world
slam shut to me
neither tasting salt of my sadness
touching skin of my joy
nor marrow of my need

Only stale scents of self-indulgence
the hollow sound of personal gain
reflect your overcast gaze

Even you do not sense your squint
against glaring loneliness
the glaucoma of self-defeat, second skin

Will you remain forever blind
eyes hunched against shared vision?

♦ JUDITH WISHNIA ♦

Pacifism and Feminism in Historical Perspective

Sometime in the early 1800s a political cartoon appeared in the newspaper of the French feminist organization, les Droits de la Femme (Women's Rights). Under the heading Universal Suffrage, a man and a woman are shown placing ballots in a ballot box. The man's ballot is marked WAR and the woman's ballot is marked PEACE. The ballot box itself rests on a stand that says, "World peace, social harmony and the well-being of humanity will only exist when women get the vote and are able to help men make the laws."[1]

This cartoon raises two questions that go to the heart of a historical analysis of women's involvement in the peace movement: the vital connections between feminism and pacifism (or more correctly, antiwar, pro-peace activity) and the almost universally accepted belief that, while men make wars, women oppose violence, and the additional implication that women are against war because they are nurturers.

In this paper I will present a (very) short history of women's involvement in the peace movements of Western Europe and the United States, concentrating on a watershed in the history of peace movements: women's activities before and during World War I. I will postulate that indeed there is a link between feminism and pacifism; most women active in the peace movements have been feminists, fighting for the political and economic rights of women as they fought against war. But I will also postulate that being a feminist did not, and does not, automatically mean opposition to war, that many feminists have succumbed to the allure of chauvinism, nationalism, and military patriotism, and that most feminists in Europe and the United States, for example, supported their country's policies in that devastating and unnecessary war of 1914–18.

Feminists who were active in peace movements did not oppose war because they were innately pacifistic, nor was it because they were socialized in their role as nurturers to be antiwar. The feminists who were antiwar shared a common political agenda—a belief in democracy, egalitarianism, and social justice. That is, they shared a political ideology, a vision of a world in which the aims of peace, feminism, and social justice were entwined.

Although we have literary examples of opposition to war throughout history, it was not until the nineteenth century, and especially the period 1870–1914, that there developed a well-organized, articulate, and political antiwar movement. Quakers, from their origins in England in the seventeenth century, were committed to peace and women's participation in their religion and society; moreover, in the aftermath of the Napoleonic Wars, the first secular peace groups were also established in New York and London.[2] In 1844 a French feminist, Eugénie Niboyet, launched the first pacifist journal to be printed in Europe, *La Paix des Deux Mondes* (soon after, *L'Avenir*). But it was not until the last third of the century—in the aftermath of the horrors of the Crimean War, the American Civil War, the Franco-Prussian War, the continuing conflict between Russia and Turkey, the ongoing imperialist conquest of Africa and Asia, and the threat of conflict between imperialist nations—that a viable peace movement was created. It was in this period of intense nationalism and imperialism that men and women of conscience became aware of the continuous danger of armed conflict, wars made even more horrendous by increasingly effective weaponry and the threat of civilian involvement. Gone were the days of limited battlefields and individual conflict. As World War I loomed closer, it became clear that, if there were to be a Europe-wide conflict, there would be millions of victims. Moreover, at a time when increasing industrialization had engendered all the ills of rapid urbanization, national budgets devoted to arms building and war devoured money that many felt should go to fulfilling the social needs of the urban and rural poor.

The last third of the century was also a time when women on both sides of the Atlantic were beginning to have a public voice. Women were active in social reform movements and the cause of women's rights. The movement for peace and international cooperation and the movement for women's rights developed at the same time. Imbued with the desire to make their voices heard in the political arena and devoted to the concept of social justice, women, most of them feminists, entered the peace movements. The majority of these women entered peace organizations dominated by male leadership, but from the beginning there were those who felt that women were different from men in their attitudes toward war and that women had a special role to play in the fight for peace. In the late 1860s in Geneva, Marie Geogg founded L'Association Internationale des Femmes, the first continental organization devoted to international peace and feminist issues. In 1870 Goegg wrote, "Men love and want war, they know that women on the other hand, do not."[3] Organizations combining the dual goals of peace and feminism proliferated. In France, Italy, Belgium, and England, the connection between women's suffrage and peace—that women would vote for peace—was made constantly. Even Bertha von Suttner, probably the most famous of European peace leaders, who worked with numerous male-dominated organizations, began to see the crucial connections between feminism and pacifism.

The notions that wars were man-made and that women were more devoted to peace than men continued to dominate feminist discourse. In the United States Julia Ward Howe, the writer of the "Battle Hymn of the Republic," made the connection between nurturant women and peace. In England Olive Schreiner wrote,

> There is perhaps no woman . . . who could look down upon a battlefield covered with slain, but the thought would rise in her, "so many mothers' sons". . . . On that day, when the woman takes her place beside the man in the governance and arrangement of external affairs of her race will also be the day that heralds the death of war as a means of arranging human differences . . . it is because on this one point and on this point almost alone, the knowledge of woman, simply as woman, is superior to that of man; she knows the history of human flesh; she knows its cost; he does not.[4]

By 1914, as millions marched in the streets, both the peace and suffrage movements had reached a fever pitch. But in August there began the conflict that was to last four long years and take millions of military and civilian lives. Feminists of belligerent nations joined their nonfeminist sisters and rallied to their respectives flags. Whether through political calculation—women would surely get the vote when they worked along with the men for victory—or whether through the patriotism that had been encouraged in an age of intense nationalism, German, French, and English women put aside their peace and suffrage activities "until victory." The Swedish-born feminist Ellen Key noted with great sadness that many of the British women who flocked to the Women's Reserve Corps were those who had marched in suffrage parades just a few months before.[5]

A small group of feminists did not give up. For them the struggles for peace and freedom from oppression were inseparable. The International Alliance for Women's Suffrage was weakened, but suffragists in these member organizations formed the nucleus of a movement to bring a negotiated end to hostilities. In the United States the stimulus for this peace activity was to come from the great social reformer Jane Addams and the suffragist leader Carrie Chapman Catt. Because of her desire for social reform, Jane Addams saw preparation for war and, of course, war itself as a tragic waste of human energy and human life. Her 1906 book, *Newer Ideals of Peace*, had established her as a respected American voice against war. Encouraged by European feminists, including the Hungarian Rosika Schwimmer and the English suffragist Emmeline Pethick-Lawrence (who already had broken with pro-war suffragist leaders, Emmeline and Crystabel Pankhurst), in January 1915 Addams and Catt organized a huge conference in Washington D.C., where representatives from numerous women's organizations, mostly suffragists, created the Women's Peace Party. Among the enthusiastic supporters of the new group were Charlotte Perkins Gilman, Fanny Garrison Villard, and Crystal Eastman.

At the end of April delegates of the Women's Peace Party joined numerous European delegates at The Hague for a huge international peace conference organized by Addams and European suffragists, including Aletta Jacobs of the Netherlands and Chrystal MacMillan of Great Britain. Socialist women had met the month before where they heard Clara Zetkin make her appeal to European socialist women to "make war on war." Not all women could make it to The Hague—it was very difficult for those in the warring countries to cross borders; others refused. The leading French feminist organizations, the Conseil National des Femmes Françaises and the Union pour le Suffrage des Femmes, wrote, "French women will bear the conflict as long as it will be necessary. At this time united with those who battle and die, they do not know how to talk of peace."[6] Yet the women at The Hague passed resolutions that would serve as a model for the hoped-for "feminist" peace. Mediation, arbitration, and international cooperation, with political rights for all, would be the basis of permanent peace. Vilified by the press, labeled traitors and hysterics (Theodore Roosevelt called them "hysterical women"), they nevertheless returned to the United States to build the Women's Peace Party and, after the war, the Women's International League for Peace and Freedom. In 1931 Jane Addams, somewhat belatedly, received the Nobel Peace Prize and in 1946, at the age of seventy-nine, Emily Greene Balch, who had been fired from Wellesley for her peace efforts in World War I, was similarly honored.

In France, on the other hand, many feminists rallied to the cause of patriotism. Indeed, as the war advanced and the tremendous casualty figures mounted, it was Marguerite de Witt Schlumberger of the Women's Suffrage Association who called on French women to do their duty after the war, to "give to the nation the children who would replace those who died."[7] But there were other feminists, many of whom were involved in socialist and union activity and many of whom were teachers, for whom there was no separation between the politics of feminism, the antiwar movement, and the politics of the working class. Their feminism was an integral part of a revolutionary political consciousness. Many of them were members of the Groupe Féministe Universitaire, a feminist group formed by members of the French teachers' union. When the war began, these women kept the union going and established contact with those socialists and unionists who opposed the continuation of the war. In July 1915 one of these feminist teachers, Marie Mayoux, edited and issued a manifesto that began with the words, "Enough spilled blood." "It was time," the statement continued, "for the Allies to end the butchery, to initiate peace talks."[8]

In the columns of the teachers' union newspaper, the women continued to combine their desire to end the war with the desire to improve the lives of women. They campaigned for the overworked and underpaid women who were employed in war factories, demanding rest periods, child care facilities, and adequate pay. They defended women teachers who had taken over the classes of the departed men and were then accused of stealing men's jobs. And always they spoke of the future. There would be no permanent and just peace, no true social change, if women were not granted their economic and political rights. In October 1916 the feminist-pacifist-socialist Hélène Brion addressed a peace group with a statement that began, "We who have been unable to do anything to stop the war because we have no civil rights are heart and soul with you in wanting to end it." In the fall of 1917 Brion was arrested for the crime of "defeatism." Accused of using her feminism to cover up her traitorous activities against the war, Brion told the court that the opposite was true; she was an enemy of war *because* she was a feminist.[9]

Other French feminists were harassed and jailed for their peace activities. The socialist Louise Saumoneau tried to lead the Socialist party into opposition to the war and appealed to women especially by organizing the Socialist Women's Action Committee for Peace and against Chauvinism. She spent months in prison. And, influenced by the meeting at The Hague in 1915, a small group of feminists organized the French section of the International Women's Committee for a Permanent Peace (Comité International des Femmes pour la Paix Permanente [CIFPP]). The leadership of this committee had also come to their antiwar activities through socialism and feminism. One of the committee's founders, Jeanne Halbwachs, was a socialist, and the other founder, Mathilde Duchêne, as prewar president of the labor section of the National Council of French Women, had initiated that organization's investigations of the evils of home work and sweatshops. Within a month of its founding, the committee issued a pamphlet, *An Urgent Duty for Women*. Describing the horrors of war, the pamphlet asked for women to demand an end to this conflict in which there could be no victors. A police raid put an end to their publications, but the committee continued to meet and work in other organizations. By 1917, Duchêne helped in organizing the League of Nations and, continuing her efforts on behalf of working women, was a leader of the Committee against the Exploitation of Women.[10]

In the United States similar conflicts over the war divided women activists. The sinking of the Lusitania and other stories of the war's atrocities brought many nonfeminist women into pro-war and preparedness organizations; once the United States entered the war, even the Women's Peace Party was hesitant to agitate for an immediate peace. The group confined its program to planning for the peace conference and for international cooperation after the war. Only the New York group, led by the radical feminist Crystal Eastman, continued to press for immediate peace. Eastman warned Jane Addams that, unless the Women's Peace Party took advantage of woman's "greater regard for life," there was no reason to maintain a separate women's peace organization.[11]

The end of the war, though it brought suffrage to English and American women—French women did not get the vote until 1946—did not bring about a feminist world of peace and justice. Nevertheless, women continued to work in peace organizations. The trauma of World War I brought millions into a movement that only faltered with the coming war against fascism, a struggle that seemed in keeping with a Feminist vision of social justice.[12] In the aftermath of World War II, with the rebirth of the feminist movement and activity against the war in Vietnam, organizations such as the Women's Strike for Peace again combined the goals of feminism, peace, and social justice. As Amy Swerdlow has noted, the women in the Women's Strike for Peace "shared the 1940's belief that society could be restructured on humanistic lines through the direct action of ordinary people."[13] Today, faced with the threat of nuclear annihilation, feminist women are once again in the forefront of the international peace movement, bringing to it their visions of a world of equality and justice.

As we have seen throughout the history of women's involvement in various peace movements, feminists and nonfeminists, with different goals in mind, have frequently supported the view that women, because they are mothers and are socialized to be nurturant, sensitive, and cooperative, are more likely to work for nonviolent solutions to problems and to be more peace-loving. Men, on the other hand, are socialized to be competitive and aggressive and are more likely to use violence and war. Recent work by feminist social scientists Jean Elshtain and Carol Gilligan support this concept of a "different voice."[14] Yet, I would postulate that women's different voice is not one that emerges from motherhood or from mere socialization—not all mothers share the same ideology, and similar socialization does not lead to similar political positions—but that it emerges from the development of a feminist consciousness that grows as part of wider political consciousness and activity. Feminism and pacifism are linked, but that linkage is based on the concept that peace, female liberation, and human liberation are inseparable.

NOTES

1. Cited in Cambridge Peace Collective, *My Country is the Whole World*(London: Pandora Press, 1984), 67.

2. The New York Group was founded in 1815, the London Peace Society in 1816.

3. For a complete history of women in the European peace movement before World War I, see Sandi Cooper, "The Work of Women in Nineteenth-Century Continental Peace Movements," *Peace and Change* 9, no. 4 (Winter 1984); and "Women's Participation in European Peace Movements: The Effort to Prevent World War I," in *Women and Peace*, edited by Ruth Roach Pierson (London: Croom Helm, 1987).

4. Olive Schreiner, *Women and Labour* (London, 1911, 1978), cited in Cambridge Collective, *My Country*, 81.

5. Cited in Lela B. Costin, "Feminism, Pacifism, Internationalism and the 1915 International Congress of Women," in *Women's and Men's Wars*, edited by Judith Stiehm (New York: Pergamon, 1983). This article previously appeared as a special issue of *Women's Studies International Forum*, 5, nos. 3/4 (1982).

6. Cited in Costin, "Feminism."

7. Archives de la Préfecture de Police de Paris, BA 1651, 20 March 1916.

8. The manifesto is in Archives Nationales, F7 13576. Other sources on teachers and the war include F7 13568, 13574, 13576, and 13743.

9. An account on the Brion trial is in the Dossier Brion at the Bibliothéque Marguerite Durand in Paris.

10. The CIFPP, Mathilde Duchêne, and their publications are discussed in dossier BA 295 *provisoire* in the Archives of the Prefecture of Police of Paris and in F7 13375 in the National Archives.

11. Cited by Barbara J. Steinson, " 'The Mother Half of Humanity': American Women in Peace and Preparedness Movements in World War I," in *Women, War and Revolution*, edited by Carol Berkin and Clara Lovett (New York: Holmes and Meier, 1980)

12. For a fuller treatment of the inter-war peace movement, see Harriet Alonso, *The Women's Peace Union and the Outlawry of War, 1921–1942* (Knoxville: University of Tennessee Press, 1989).

13. Amy Swerdlow, "Ladies Day at the Capital: Women's Strike for Peace versus HUAC," *Feminist Studies* 8, no. 3 (Fall 1982).

14. See Jean Elshtain, "On Beautiful Souls, Just Warriors and Feminist Consciousness," in *Women's and Men's Wars*, edited by Judith Stiehm; see also Elshtain, *Public Man, Private Woman* (Princeton: Princeton University Press, 1981) and Carol Gilligan, *In a Different Voice* (Cambridge: Harvard University Press, 1983).

♦ SALLY L. KITCH ♦

Does War Have Gender?

In her post-Vietnam novel, *In Country*, Bobbie Ann Mason poses two questions: does war have a specifically male or masculine character?, and, if women were in power, would war exist? Samantha, the teenage female character who muses upon those questions, presents, almost at once, two seemingly contradictory answers. First, after reading the horrifying diary of her dead father's last days in combat during the war, Samantha observes, "If it were up to women, there wouldn't be any war. No, that was a naive thought. When women got power, they were just like men. She thought of Indira Gandhi and Margaret Thatcher" (208). Just a few moments later, however, she posits the apparent opposite of that conclusion: "Women didn't kill. That was why her mother wouldn't honor the flag, or honor the dead. Honoring the dead meant honoring the cause" (210).

Clearly Samantha is a mixed-up girl. Or is she? A closer look at the terms in which she thinks about women and war reveals that her confusion may reflect her failure to distinguish between physiological sex and culturally defined and sanctioned, sex-related but not genetically determined behaviors and attitudes. Such culturally defined characteristics serve as a kind of gender ethos for each sex that roughly parallels notions of masculinity and femininity. Because she confounds the physiological and cultural categories, Samantha's understanding of the proposition before her—that men are responsible for the existence of war—is instantly contradicted by an anomaly. How can she explain Margaret Thatcher and Indira Gandhi? They are physiologically female, a fact that seems to imply a rejection of war and its symbols, and yet they have been responsible for war and all its horrors. Are they somehow male? How can they coexist in the same world as her mother and other ordinary women who seem, in their disgust with war, to represent a kind of female norm?

The answers to Samantha's questions lie in the cultural gender ethos to which her anomalous figures have attached themselves. Even to Thatcher's supporters, her policy in the Falkland Islands (the war behavior Samantha is thinking about) did not enhance her reputation for femininity. In fact, much of the public praise she received emphasized her toughness and resolve, her ability to overcome human considerations and protect British territory just as a male, or "regular," politician would do. Obviously, nothing in her physiology prevented her behavior, but Samantha senses that her actions were not consistent with the female gender ethos. Her policies have, indeed, masculinized her, though she remains a female.

In light of Samantha's confusion, and of other recent debates about man as war maker and woman as peacemaker, we must emphasize not the physiological sex of the politician or soldier but the existence, at least in the Western world, of a gendered ethos that informs our understanding of peace and war. We must also consider the proposition that values and worldviews contained within the personal gender ethos have public, political implications. In fact, a structural approach to war requires us to see the connections between personal and political issues, including the issues of sex and gender.

This article will suggest that war has gender, rather than sex, by demonstrating the connection between the war ethos, which is a matter of public policy and political behavior, and the ethos of masculinity, which relates to qualities of personal identity. While the male ethos defines a highly individuated, autonomous, unemotional, rational, and powerful human being, the war ethos promotes competition, authoritarianism, and the use of coercive force to maintain or promote a social system often deemed self-evidently good (Reardon 10). Among the many examples of the linkage between masculinity and war is the fact that participation in one (war) historically has served to enhance the other (masculinity), even in a woman like Margaret Thatcher.

THE PERSONAL AND THE POLITICAL: ORIGINS OF MASCULINITY AND WAR

Also among the shared characteristics of the male gender ethos and the war ethos is an emphasis on a well-defined sense of self, individual autonomy, and self-reliance. The emphasis establishes a "we-they" understanding of relationships and promotes competition in which there are clear winners and losers. War requires a view of another nation or race of people as an enemy with whom we share little and whom, therefore, a soldier can hate and kill with honor. Masculinity similarly depends in large part on the distinctions between people, both among and between men and women. The male gender ethos is often defined in terms of qualities that are most clearly not-female or nonfeminine. The insult to a boy implicit in the taunt "You're acting like a girl" appears obvious in our culture. Men can frequently be dissuaded from performing a task by being told that it is really women's work. Men learn to assume that real men are the opposite of real women, and they worry about men who seem not to be sufficiently differentiated from women.

The male ethos and the war ethos also share an emphasis on superiority and dominance as the logical conclusion of difference. In warfare, the fact of difference between two sides in a conflict implies the proposition that one is superior to the other; in war each side will accept nothing short of victory. In the male gender ethos, men assume their superiority over and right to dominate those who are not-men. To be "different from" the apparent norm is to be "subordinate to" in both systems of thought.

The female gender ethos focuses on other issues. Difference from the male is of less concern; calling a girl a tomboy is not especially insulting. The female gender ethos partakes primarily of the tradition of women's presumably natural nurturing role and the lack of social status that role has historically entailed. Thus, the female gender ethos emphasizes self-sacrifice, putting others' needs before one's own, the maintenance of human relationships, and the preservation of fragile life in a hostile world (Ruddick 216–18). On a public level, such a personal ethos tends to support nonviolence, conciliation, and attention to issues concerning the quality of human life.

The processes involved in the internalization of a culturally defined gender ethos are sufficiently complex that they have often been mistaken for genetic or hormonal determinism. Gender differences in behavior have been detected in very young children, for example, and such differences seem difficult to explain in terms of learning. Yet the behavior of women like Margaret Thatcher and men like Linus Pauling and Alan Alda, not to mention the apparent sex-role reversals found in the anthropological work of Margaret Mead and others, are impossible to explain in hormonal or genetic terms.

Differentiated gender norms in behavior and attitude, even among children, are explained in the work of such feminist scholars as Nancy Chodorow and Carol Gilligan. Both explain that the learning of a gender ethos is embedded in complex psychosexual issues that are, in turn, culturally determined. Central to the theories of Chodorow and Gilligan (who bases her work on Chodorow) is the fact that women do the child-rearing in Western cultures. Given the emphasis on individualism in such cultures, child development is dominated by the need to separate and individuate from the mother. Children of both sexes go through this process, but for boys the task of separation is more intense; they must sever ties to mother completely if they are to earn a masculine identity. This process promotes misogyny among men. Girls also learn to separate and, in a misogynist society, to disparage their mothers. But the similarity of physiological sex between mother and daughter makes an ongoing connection between the two less threatening and contributes to the development of a culturally approved feminine ethos (Chodorow 168–69). Central to the applicability of this scenario to the topic of war is the fact that inner constructs of the human psyche are linked to social and political structures (Reardon 6). In male-dominated cultures, individuated male, who learns misogyny as a feature of his own identity, is more likely than the female to become the politician or general who determines foreign and domestic policy.

Gilligan applies the culturally determined differences in gender ethos to the social level of moral decision making. She claims that boys develop abstract, analytical approaches to ethical questions in the process of separating from the mother, while girls develop “an ethic of care” in which the maintenance of relationships, a feature of their more elastic, continuing connection with the mother, tempers the completely abstract consideration of moral questions (Gilligan 73–105).

Gender differences that parallel the psychosexual concerns of male and female children are manifested in their play. Because both individual needs and cultural expectations determine the appropriateness of games and toys for children of each sex, patterns of play, like patterns of moral decision making, must be seen as both reflections and creators of the dual-gender ethos system of our culture. Janet Lever’s studies of children’s games reveal that boys are more likely than girls to play team sports in which winning is emphasized. Such games require a “we” opposed to a “they” who must be beaten. Girls, on the other hand, learn to play games that require taking turns, such as jump rope, and de-emphasize a specific outcome or victory. When conflicts develop in boys' games, the need for a victory encourages them to settle their difficulties by specified rules; boys are socialized to be more interested in a game's results than in their friendships with the other players. Girls are more likely than boys to abandon their play in order to preserve relationships among the players. Girls try to mediate and resolve conflicts rather than to devise rules for their resolution (Lever 67–72).

Other studies have suggested that boys learn early to focus on objects and tools in problem solving, a focus that might direct them toward technology rather than interpersonal interaction in their social and political behavior. As technology acquires masculine overtones, its use appears to promote masculine values, and masculine values appear to require technological approaches. The male gender ethos, therefore, emphasizes rationality rather than emotion, "hard" rather than "soft" solutions (results rather than process), and quantitative rather than qualitative measures (see, for example, Caldicott 296–97).

Because both the male and female gender ethos are learned, the exact correspondence between physiological sex and ethos can go awry. For example, a girl can blend characteristics of each ethos or internalize the male system of attitudes, values and beliefs. A study of Margaret Thatcher's family and school experiences might reveal an unusual set of circumstances for a girl and, therefore, the origins of her more masculine ethos. Alternatively, her successful entry into high levels of politics in a Western nation might have caused her masculine bent. Thatcher undoubtedly had to demonstrate values and skills that were acceptable to the males who dominated the political system. Physiological women can, of course, easily comply with such a requirement, given the opportunity. Males can also adopt and internalize values of the female ethos.

GENDER AND THE SOCIAL MECHANISMS OF PEACE AND WAR

An examination of the political arena reveals several levels on which peace and war have acquired, respectively, feminine and masculine connotations. At the simplest level, groups dedicated to peace and arms control may include members of both sexes, but only the female gender ethos is associated with peace activism in the titling of such groups. The most important index of organizations (the Encyclopedia of Organizations) listing dozens of peace and arms control groups includes no groups with "men" or "fathers" in their titles, although Albert Einstein and Harold Stassen have peace organizations named for them. The same source lists ten groups with "women," "mother," or a particular woman's name in their titles, including Another Mother for Peace, Women Against Nuclear War, Women Strike for Peace, and the Women's International League for Peace and Freedom. Europe offers Irish Peace Women, Women for Peace (Nordic), and the Women of Greenham Common. Peace groups with feminine identities also exist in Japan, Australia, and Argentina. (When men decide to emphasize their gender in organizations, they tend to do so in religious groups or clothing associations. There is also a Society for Dirty Old Men and an organization of American War Dads [*Encyclopedia* passim]). Groups of men or mixed-sex peace groups focus on peace not as an aspect of male gender ethos, or of gender or sex roles at all, but rather as an aspect of particular occupations. Examples of such groups include Scientists Against the Arms Race, Writers for Peace, Lawyers' Alliance for Arms Control, and Athletes United for Peace. Men might regard peace work as socially and professionally responsible, but they do not associate it with masculinity. Indeed, whatever connotations of masculinity certain occupations may have are deemphasized in the names of these organizations.

Military jargon presents the opposite picture. Terms used to describe and define weapons and weapon delivery systems often have quite explicit connotations that support the connection of male sexuality with war and other forms of violence. In her book, suggestively entitled *Missile Envy*, Helen Caldicott cites several examples of such terminology, including *missile erector, deep penetration*, and *soft laydown*. Whether or not the weapons are phallic, they are perceived to be phallic by their designers and users. Caldicott seems to argue for a genetically determined male proclivity for war, but such a claim need not be supported for the cultural connection between weapons and images of male sexuality to become obvious. Caldicott also recalls that General George Patton defined war as the "cataclysmic ecstasy of violence" and that the chair of the Committee on Public Information during World War I believed that "universal military training means . . . national virility" (Caldicott 297–98).

Some might argue that the presence of women in the military contradicts the existence of a peace-identified feminine ethos. The circumstances of women's military participation may, however, actually reinforce the dissociation of the female ethos from war. First, although women serve in the military, certain military functions, including armed combat, are considered off-limits to women. This attitude persists despite the fact that women have performed combat service in several recent military operations.

Although there are a variety of arguments against women in combat, it is clear that one deep reason for the prohibition is the contradiction some politicians and military men perceive between the male ethos of combat and the presence of females in the war zone. Senator Sam Ervin's 1972 testimony against the Equal Rights Amendment (ERA) provides one example of the perceived contradiction. Ervin opposed the ERA partly because he opposed its association with the drafting of women into military service (a possibility that exists under the Constitution even without the ERA). In addition to his belief that women draftees would procreate like bunnies without paternal control of the progeny, Ervin feared that a basic rationale for war—a man's desire to protect the women at home—would be destroyed by the presence of women soldiers. "I am satisfied that the veterans who waded in the icy waters in trenches until their feet bled during the First World War . . . who endured the heat of North Africa and the Anzio beachhead . . . during the Second World War, and the veterans who fought in the mountains of Korea and the steaming swamps of . . . Vietnam are implacably opposed" to subjecting American girls to such conditions (Ervin 23). Men cannot be men if they cannot protect women. If girls, too, were in the swamps and jungles, what would the boys be fighting to preserve?

A more recent slant on the importance of gender difference to the war ethos was expressed by a writer for the *Air Force Times*. He wondered "whether women are capable of the explosive savagery required by combat." He was not worried about men who, "a quarter-inch below the surface, are dismal animals," but women are another story (Reed). They might not be reliable in the crunch.

A closer examination of women's roles in the military around the world further supports differences in male and female gender ethos relative to war. Birgit Brock-Utne claims that women's military roles reflect their civilian femininity rather than promote their gender neutralization or even their masculinization. By and large (even in Israel, where military service is mandatory for women) in the armed forces women cook, serve, type, and manage offices. Thus, they support male war efforts without fully participating in them. Women's exclusion from combat also limits their access to the most upwardly mobile jobs in the military, which keeps women out of policy and decision-making positions in war matters, just as they are kept out of civilian politics. In fact, Brock-Utne insists that women are considered valuable in the service primarily because their enculturation in the female gender ethos makes them malleable, docile, and diligent. Even with child care responsibilities, women's rate of absenteeism in the armed forces is one-half that of men. Women are also less likely than men to use narcotics or cause disciplinary problems (24–25).

In general, Brock-Utne deplores female participation in the military as an exploitation of the female gender ethos by the masculine war machine. She suggests that women must "explore how they are utilized by male-run states and why. Otherwise they risk labeling as progress what in reality is an adjustment to the state elite's own latest manpower imperatives" (23). American women's experience in Saudi Arabia during the Persian Gulf crisis is a case in point. American military women were required by their government to bend to the discriminatory and regressive customs of their Arab hosts. Such restrictions undermined the legitimacy of women's achievement of power as women, as well as the government's support of that achievement.

The fact that women's presence in the military has not resulted in the incorporation of female gender ethos values in the policy or practice of war reinforces the association of masculinity (as a cultural construct) with war and suggests other implications of women's exclusion from the battlefield. A genuinely mixed-sex military at both the combat and the decision-making levels might interfere with whatever sexual meaning men derive from their dominance of weapons systems and even, as Caldicott suggests, with killing (296). It might also violate the male dominance of the moral imperatives that derive from war. If only the warrior can be considered courageous, then woman's (happy or sad) exclusion from combat denies her that honor. The moral hegemony of the war system discounts women's bravery in childbirth or any other feminine role. That system also grants the soldier the privilege (and responsibility) of making decisions about life and death on his own authority. By denying women the warfare role, the war system denies them the moral authority to make such decisions. That denial extends to women's other social roles, even their childbearing function, which, though unique to women, is not under their moral control. Rather, women's moral choices about childbearing, such as the decision for abortion, have been usurped by the (male) state, which presumably earns its authority over women through its control of war. Finally, as comrades in arms, women would threaten the typical identification of their sex as spoils of war; they would lose their status as objects to be won and used.

THE MALE GENDER ETHOS AND THE POLITICS OF WAR

The military is not the only arena in which war is defined and waged. In the political arena, too, where war and peace policies are formulated, the male gender ethos is also connected with the ethos of war. Like military structures, the political structures that support war can also be seen as masculine, according to Jean Bethke Elshtain (52). She argues that the concept of nationhood itself is masculine, because a nation is defined as an independent state that bows to no international rule. Nations are famous for taking pride in their sovereignty and autonomy, as men are, and they prefer international anarchy to the subordination of their interests to global concerns. In such a world, as on the boys' playground, conflict is almost guaranteed (Elshtain 40).

Also in such a world, conflict resolutions that depend upon cooperation, negotiation, and compromise are frequently perceived as feminine and are, therefore, disparaged. The United Nations, for example, has been ignored by powerful nations, including the United States, when its requirements of compromise and deliberation have appeared threatening to national sovereignty. The disparagement of the UN by the United States has been accompanied by the appointment of several women—from Eleanor Roosevelt to Jean Kirkpatrick—to the relatively powerless position of ambassador to the UN. In fact, women have apparently been more acceptable in the role of UN ambassador than in most other American political roles. In the context of selective respect for that international body by recent American presidents, such acceptance of the feminine only reinforces the marginality of the UN's form of peacemaking in American culture.

Another indication that contemporary war and peace policies are related to the issue of gender ethos is the typical separation in American culture of public issues from concerns labeled private or personal. Attempts to infuse the public world with private (feminine) issues or models are generally resisted as naive or as a means of trivializing important social concerns (Elshtain 41). Except at election time, male leaders are seldom identified as fathers, and they rarely discuss political issues in terms of their fatherhood roles. What is defined as realpolitik is the masculine/public rather than the feminine/private side of the human ledger (Elshtain 43–45). The exclusion of the personal from the political also means the exclusion of what is seen as the soft side of human experience—low in technology, qualitative, and fuzzy-headed—from the politics and ethos of war. The concomitant separation of human beings into (male) warriors and (female) mothers carries the ethos of war into all arenas in which men dominate (Reardon 51). Consequently, any wisdom and knowledge endemic to female-dominated activities and the female gender ethos are eliminated from politics even, as we have seen, when women get into power through male-defined channels.

Betty Reardon has suggested that the sexist exclusion of private/feminine elements from politics, as well as other forms of sexism, reinforce the viability of war in our society by promoting the values of hierarchy and competition and supporting the necessity of dominance and subordination in the presence of difference. Sexism in politics—defined as oppression in the achievement and maintenance of power—parallels the choice of violence (war) as a political solution (Reardon 54–55). War is a logical outcome of a politics that promotes competition and disdains family and personal relationships as political models.

Feminists have suggested that the infusion of the male gender ethos into politics has become increasingly dangerous. The United States discovered in Vietnam, and may discover in the Middle East that "power-over" solutions are not necessarily the most effective because they presuppose a level of control that may not be possible. The power-over approach is also limited in the modern world because its primary assumption—that beneficial change follows apocalypse—becomes moot in the nuclear age (Elshtain 51). New beginnings are unlikely to come from the ashes of a nuclear war.

Although the parallels between modern political structures and the male gender ethos are striking, the masculinity of such structures has generally been hidden from our collective consciousness. The male gender ethos itself encourages such masking because it teaches that its characteristics define the human condition and makes of the not-male a subordinate other. Women learn these rules as well, since the female gender ethos teaches self-deprecation and subordination as positive feminine qualities. Even women who resist the presumptions of the male gender ethos might deny what they see and feel for survival purposes. People successfully socialized for dependence are unlikely to challenge authority, even if it is illegitimate.

Yet, in a culture as highly gendered as our own, true gender neutrality is rare at any social level. Specific gender values are particularly likely in realms, such as politics, that have been blatantly dominated by one sex. In such arenas, there have been too few opportunities for the balancing or neutralizing values of the female gender ethos to be introduced. Similarly, male ethos values (and accompanying material rewards) have been absent from female-dominated realms, such as homemaking, the helping professions, and child care.

FEMALE GENDER ETHOS ALTERNATIVES

The important question, as we have seen, is not what women would do if they were in power in the current political climate but, rather, what the female gender ethos might contribute to the politics of war and peace if it were taken seriously. There is evidence that the female gender ethos is more comfortable with birth and nurture than with apocalypse; it conceptualizes change as a slow, evolutionary process, and it encourages the maintenance of human relationships over the achievement of specified results. Definitions of power in the female ethos relate to the female "epistemic and social location" in motherhood, which requires a concept of power through restraint (*ascesis*) and depends on the possibility of forgiveness (Elshtain 53). The maintenance of life opposes violence; nurture requires the continual refusal of the nurturer to crush the fragile being in her care. If mothers defined their power as nations do, few children would survive.

The key question is whether politicians, male and female, can learn to define political power in maternal terms. Such learning seems to be a timely goal, since for the first time in history human beings have now attained the power, once reserved for God, to destroy not just other nations but the entire human race. Before the bomb, believers gave to God their obedience and faith in order to ensure the restraint of that absolute power. Because of the bomb, the female gender ethos might be the only means of ensuring humanity's exercise of that Godlike restraint.

The connection of warfare to gender ethos values recalls a common cliché of American culture, which characterizes the male-female relationship as a "war between the sexes." That metaphor connects gender difference with the war mentality by associating difference with conflict. The metaphor also offers a possible opening into the morass of the peace and war question in modern society, however, if we accept the premise that political solutions can begin with personal issues. The elimination of the interpersonal gender war requires a reduced focus on difference as opposition and hierarchy. Similarly, the elimination of war may depend on a reduced focus on national differences and an increased emphasis on human connection.

Placing women in charge of extant political structures, inextricably bound as they are with war making, will serve little purpose. For change to occur, the political world must be infused with female gender ethos values that protect life, accept diversity without hierarchy, and recognize the destructiveness, rather than the sexiness or technological elegance, of weapons. The female gender ethos values the quality of life above the quantity of warheads, and it recoils at risking global destruction to protect ownership of territory or resources. As Reardon suggests, the female gender ethos may be the most truly international value system on the planet (68). The needs of nurturance are universal; they transcend political boundaries and offer an opportunity for unity despite human conflicts. The female gender ethos, therefore, can provide the grounds for a new international understanding and, possibly, the first true commitment to global peace in the history of the world.

WORKS CITED

Brock-Utne, Birgit. *Educating for Peace: A Feminist Perspective*. New York: Pergamon Press, 1985.

Caldicott, Helen. *Missile Envy*. New York: William Morrow, 1984.

Chodorow, Nancy. *The Reproduction of Mothering: Psychoanalysis and the Sociology of Gender*. Berkeley: University of California Press, 1978.

Elshtain, Jean Bethke. "Reflections on War and Political Discourse: Realism, Just War, and Feminism in a Nuclear Age." *Political Theory* 13 (February 1985): 39–57.

Encyclopedia of Associations, 19th ed. Detroit: Gale Research, 1985.

Ervin, Samuel. *Congressional Record* (28 March 1972): 10450–56; reprinted in Organization of American Historians *Newsletter* (August 1983): 23.

Gilligan, Carol. *In a Different Voice*. Cambridge: Harvard University Press, 1982.

Lever, Janet. "Sex Differences in the Complexity of Children's Play and Games." In *Feminist Frontiers*, ed. Laurel Richardson and Verta Taylor, 63–73. Reading, Mass.: Addison-Wesley, 1983.

Mason, Bobby Ann. *In Country*. New York: Harper and Row, 1985.

Reardon, Betty. *Sexism and the War System*. New York: Teacher's College Press, 1985.

Reed, Fred. "Women in Combat: A Real Bad Idea." *Air Force Times* (29 January 1990): 70.

Ruddick, Sara. "Maternal Thinking." In *Mothering: Essays in Feminist Theory*, ed. Joyce Trebilcot, 213–30. Totowa, N.J.: Rowman & Allanheld.

♦ MARY BROWN PARLEE ♦

Women, Peace, and the Reproduction of Gender

The purpose of this essay is to consider from a psychological perspective some of the ways gender is related to peace and justice. By a psychological perspective, I mean one that focuses on actions, feelings, and thoughts of individuals who are located in both cultural and biological contexts and who cannot be described or understood apart from them.

Certain questions frequently arise when intelligent people talk about psychological issues related to women and men, peace and war. They appear to be questions that have or could have scientific answers—for example, are women less aggressive than men? If they are, is it because the "male" hormone, testosterone, causes males to be more aggressive? Or is it that males and females are socialized differently so that there is a gender difference in aggression? Are women more nurturant, empathic, and caring than men? If they are, how are such gender differences shaped by biological and social processes? Are there gender differences in traits other than aggression and nurturance that would lead women, if they could, to organize society so that war would not be the ever-present possibility it is in a world organized (in the public sphere) almost exclusively by men? Could the threat of war be reduced or eliminated by changing the way males and females are raised to be "masculine" and "feminine" (as these are defined in our culture)? The assumption underlying such questions, of course, is that people who are aggressive might be more likely to think of and use force as a way of resolving competing interests, while nurturant people might see and prefer different resolutions to problems—including redefining the situation from one of competing rights and interests to one of interwoven relationships and responsibilities for common concerns (Gilligan 1982).

While these may seem to be interesting and relevant questions, the argument offered here is that they are not the right ones for psychology as a scientific enterprise nor for intelligent, concerned citizens to ask or try to answer. Among the reasons they are not, two are most important. One is that they implicitly endorse generalizations about "women" and "men" as the appropriate language for thinking about gender. The other is that they depend on an analysis of the human personality in terms of "traits," and they implicitly present this approach as the only kind of psychological analysis possible. These assumptions are not only wrong in themselves, but in their surface plausibility they serve to keep our thinking directed away from alternative kinds of psychological analyses and the psychological research and social actions for peace that might flow from them.

As they are used in practice, the concepts and methods of research psychologists typically focus on the actions, thoughts, and feelings of individuals stripped from contexts (Mishler 1979; Parlee 1979). That is, they implicitly represent persons as autonomous, rational individuals, each interchangeable with everyone else for theoretical purposes. Such "context stripping" is compatible with, and very frequently leads to, psychological questions posed in terms of the traits of individuals and generalizations about males and females. This approach has also led many psychologists, like many nonpsychologists, to ask the question (and others like it), Are women or men more aggressive?

Seeking an empirical answer to this question, Eleanore Maccoby and Carol Jacklin exhaustively surveyed the research literature a decade ago and concluded that the weight of evidence showed that boys and men are more aggressive than girls and women (Maccoby and Jacklin 1974). Although some psychologists have disagreed with Maccoby and Jacklin's interpretation of the data (Block 1976), few have argued with the basic assumptions that traits like aggression are the appropriate units of psychological analysis and that generalizations can be discovered about males and females. These assumptions are embodied not only in most of the psychological research on gender differences, but also in everyday discourse whenever questions are asked about differences between females and males.

As both scientists and feminists, however, feminist psychologists need to go beyond seeking generalizations in psychological theories about "women" and "men" or "males" and "females." The empirical evidence is now overwhelming: there simply are no such generalizations one honestly can make, even about phenomena widely shared within one gender and not the other, such as menstruation or devaluation within dominant cultural ideologies (Parlee 1981). Thus, general statements that universalize male and female (thereby implicitly stripping women and men from their biological and social contexts) do not make scientific sense in part because they lead to descriptions and conclusions that are empirically false. When such generalizations are used in psychology or in everyday discourse, they usually support or refute an ideological claim about gender rather than promote scientific understanding. (Shields has persuasively documented this phenomenon in research on gender differences in brain functioning [1975].)

As Jessie Bernard pointed out many years ago, at most, scientists can hope to discover generalizations that are empirically true of "some women" and "some men" (Bernard 1974). And then, of course, the research is concerned with discovering, for example, which men are more aggressive than which women—and the whole direction of thinking shifts slightly but significantly. In this way, scientists move from asking descriptive psychological questions about a dichotomous classification of persons into male and female (which is not an empirically adequate description even at the biological level) to the beginnings of explanatory questions about biological, social, and psychological processes. For example, they ask, Is testosterone level (regardless of whether in male or female) associated with aggression? Is playing with dolls (by girls or boys) associated with nurturance? How are such relationships, if they exist, influenced by the fact that the cultural dichotomy of gender means that aggression and doll play have different social and psychological meanings for females and males? It is important to note that the psychological processes to which these questions point can be described and understood in important respects without reference to the biological sex of individuals. Asking general questions about males and females, men and women (where "all men" or "all women" is implicit) not only serves to obscure the importance of questions about processes. It also serves to obscure the very substantial ways in which class, ethnic background, education, and a whole host of other social experiences result in differences and similarities among some women and some men.

Thus, when generalizations are made about sex and gender, the questions for the social scientist and the concerned layperson or activist always should be: *which* women? and *which* men? And when general questions about (all) males and (all) females are rejected, questions about "which" and "some" immediately point to differences in processes that might be involved in producing behavioral differences. It certainly does not make scientific sense to ask if men are more aggressive than women, and it probably does not make sense to ask why some men are more aggressive than some women, since male and female might not be concepts referring to explanatory processes. The more sensible (and scientifically answerable) questions might be, Why are individuals with high testosterone levels more aggressive? If some women are less aggressive than some men, which individuals do we think they are? Are they to be identified in terms of hormone levels, socialization experiences, or combinations of these? (How) does it matter whether the hormone levels or socialization experiences are associated with being male of female? These are the genuinely scientific questions—questions about the processes implied by a more precise specification of individuals in their biological and social contexts.

The ideologies of gender that support existing social arrangements are couched in terms of men and women, and, by implication, all men and all women, but psychological science need not accept that language—(nor the concepts it implies) or use methods that permit this language to remain unchallenged. And psychologists should not accept these limitations if their task is to develop research and theory that will enable us to analyze and reveal the ideological formulation for what it is. The task will certainly not be accomplished if, instead of unmasking ideological assertions, psychologists focus on trying to prove or disprove these claims "scientifically."

A psychological analysis that does not accept the context stripping concepts and methods of mainstream psychology and everyday language and thought could be very useful in developing a challenge to the dominant cultural ideologies about gender, particularly those ideologies that support social institutions which seem to be moving us to war rather than to peace. An important part of the task of individuals concerned with social justice and peace is to reject simplistic talk and thinking about gender, both within psychology and in more widely heard public arenas. Certainly they must reject pseudoscientific research based on questions embedded in masculinist ideology.

A reformulation of thought and talk and research is needed not only for the concepts women and men, male and female, but also for conceptions of the individual in terms of supposed traits. Again consider aggression as an example of the kind of concept requiring critical analysis. If a psychologically meaningful definition of aggression involves at least an intention to hurt another person and actions to carry out this intention, then aggressive acts are defined in part in terms of the opportunities afforded by particular settings. A person with access to material and ideological resources can hurt others in ways not available to someone without that kind of power. A physically strong person can hurt or intimidate the less strong, but not vice versa. An individual with the power to withdraw or withhold love in a close and important personal relationship can hurt the person who is dependent on the relationship but who does not have the same control over the emotional transactions that determine its quality. To the extent that they spend time in different kinds of settings, differences among individuals might be related to the kinds of aggressive acts they perform rather than to aggression, as something meaningful apart from settings. The question is not, Is this person aggressive? It is, rather, How can an intention to hurt others be manifested in this setting, and is this person likely to do it?

To the extent that particular situations include more men than women as participants—say, a meeting of top officials in the U.S. Department of Defense—then the opportunities to act aggressively that are afforded by that setting will be related to gender, but perhaps only indirectly and noncausally. Similarly, to the extent that other situations include more women than men as participants—say, park bench conversations at the playground that focus competitively on children's accomplishments—the opportunities to be aggressive afforded by that setting might be related to gender. Again, however, the relationship between the aggressive actions possible in this setting and the gender of those who carry them out is likely to be indirect and noncausal.

General questions about aggression or aggressive individuals, questions formulated without specifying the contexts within which aggressive acts necessarily occur, tend to obscure the importance of the setting and the opportunities it affords for an individual to hurt others in particular ways. Psychologists and others need to focus on precisely these situational contexts, however, if they are to ask questions that are relevant to understanding psychological issues related to peace and war. For example, it is important to be clear about the fact that, in the most immediate and simple sense, wars are launched and directed from the White House, Congress, and the Defense Department, not from the park bench or classroom or secretarial desk. This is not to say that settings in which more women than men are found (in the United States in 1990) are not important and influential in shaping the thoughts, feelings, and actions of those who most immediately decide and carry out actions related to war and peace. But it does point to the way in which the concept of aggression is inextricably related to settings and the opportunities they provide for hurting others. The same analysis—that is, a focus on persons in their specific settings rather than on traits in the heads of isolated individuals—also applies, of course, to other supposedly general traits like nurturance, empathy, and caring. In short, the scientific question is not whether some (group of) people are more aggressive or more nurturant than others, but how likely are these people to act in a particular way in this setting given the opportunities it affords—for example, to hurt or care for others. Such a reformulation, an alternative psychological analysis to implicit trait theories, suggests that changing the setting might be the most effective means of changing the actions of individuals, which of course has implications for social action. Perhaps having a woman in the White House—with all the opportunities for action the presidency offers and precludes—would not make much difference.

Individuals certainly do seem to differ in their more or less stable dispositions to act with particular intentions: some people often seem to intend to hurt others and do so using whatever opportunities are afforded by the situations in which they find themselves, while others do not. Given that research questions have traditionally been formulated in terms of behaviorally defined traits rather than in terms of intentions expressed variously in different settings, there is at present little systematic empirical evidence of what leads to stable individual differences in personality (described in terms of intentions or goals) and whether or not such differences are related to sex or gender. These considerations suggest a set of complex but potentially answerable psychological questions, and both scientists and citizens would do well to insist on their complexity and to reject the simplistic trait formulations that implicitly ignore social contexts. Like generalizations about women and men, they appear to serve primarily an ideological function.

A final critical point about context stripping methods and concepts is that they seem to have promoted or to be consistent with a scientifically unworkable conceptualization of biological and social "influences" (on, for example, supposedly gender-related psychological phenomena). The conceptualization is of separate, independent processes—the biological and the social—which "interact" with each other to produce, at the psychological level, the thoughts, feelings, and actions of individuals. Since it is an empirical fact that biological and social phenomena always co-occur as aspects of a larger system, stripping them from their contexts and representing them as independent causal processes is demonstrably false and has been scientifically unproductive. It might prove easier to reject or go beyond simplistic, ideology-based questions about the origins of psychological phenomena (especially questions that appear to pose a biology-socialization dichotomy) if we reject misleading generalizations and insist on knowing *which* women, *which* men, and *what* settings are being discussed.

Despite the predominance of trait analyses and generalizations about males and females in research about gender, there is beginning to emerge in psychology an alternative, more sophisticated analysis. The strands forming this new perspective, which has informed the critique offered here, are many. They include ideas and emphases from activity theory, with its focus on cultural origins of thinking; from feminist work combining psychoanalysis with a Marxist sociological perspective (Chodorow 1978; Harding 1986); and from the theoretical account of persons in relation to social structure and ideology developed by European social psychologists such as Tajfel (1981), Shotter (1984), and Harre (1984). While diverse in subject matter and theoretical concepts, these research traditions can be seen as sharing a common central concern: analysis of the ways social structure and culture are reproduced and maintained through the psychological structure and functioning of biologically distinct and unique individuals. At present, some of the most detailed empirical and theoretical work being done in this spirit is exemplified in multidisciplinary research on culture and depression (Kleinman and Good 1985).

In time and under auspicious circumstances, it is possible that such work will develop into a coherent body of concepts, methods, and theory that will enable us to understand the systematic ways in which settings, bodies, psychological histories encoded in personality, and ideology come together in the construction of persons who have the capacities—thoughts, feelings, and actions—to make peace rather than war. It seems likely that such persons will be both female and male, but that we will not know unless we can reject questions that limit us to an examination of ideological representations of gender rather than lead to an understanding of processes that have culture-specific, gender-related psychological outcomes. The successful integration of this new approach into the academic discipline of psychology depends on whether the primary function of psychology is to provide support for the dominant cultural ideologies of gender or whether it can inform popular thinking about gender and serve as a source of alternative, critical, and liberatory knowledge, promoting social change, justice, and peace.

NOTES

1. The suggestion outlined here is similar to developmental psychologists' recognition that chronological age, per se, is not a variable in an explanatory theory and is not necessary or useful in explanations of developmental phenomena once the relevant (partially correlated) causal processes have been identified.

REFERENCES

Bernard, J. [1974] 1976. *Sex differences: An overview.* New York: MSS Modular Publications. Module 26. Reprint. *Beyond sex-role stereotypes: Readings toward a psychology of androgyny*, edited by A. G. Kaplan and J. P. Bean. Boston: Little, Brown.

Block, J. H. 1976. Issues, problems and pitfalls in assessing sex differences: A critical review of *The psychology of sex differences. Merrill-Palmer Quarterly* 22:283–308.

Chodorow, N. 1978. *The reproduction of mothering: Psychoanalysis and the sociology of gender.* Berkeley: University of California Press.

Gilligan, C. 1982. *In a different voice: Psychological theory and women's development.* Cambridge, Mass.: Harvard University Press.

Harding, S. 1986. The instability of the analytical categories of feminist theory. *Signs: Journal of Women in Culture and Society* 11:645–64.

Harre, R. 1984. *Personal being: A theory for individual psychology.* Cambridge, Mass.: Harvard University Press.

Kleinman, A., and B. Good. 1985. *Culture and depression: Studies in the anthropological and cross-cultural psychiatry of affect and disorder.* Berkeley: University of California Press.

Maccoby, E. E., and C. N. Jacklin. 1974. *The psychology of sex differences.* Stanford, Calif.: Stanford University Press.

Mishler, E. G. 1979. Meaning in context: Is there any other kind? *Harvard Educational Review* 49:1–19.

Parlee, M. B. 1979. Psychology and women: Review essay. *Signs: Journal of Women in Culture and Society* 5:121–133.

Parlee, M. B. 1981. Gaps in behavioral research on the menstrual cycle. In *The menstrual cycle.* Vol. 2, *Research and implications for women's health*, edited by P. Komnenich, M. McSweeney, J. A. Noack, and S. N. Elder. New York: Springer.

Shields, S. A. 1975. Functionalism, Darwinism, and the psychology of women: A study in social myth. *American Psychologist* 30:739–54.

Shotter, J. 1984. *Social accountability and selfhood.* Oxford: Blackwell.

Tajfel, H. 1981. *Human groups and social categories: Studies in social psychology.* Cambridge: Cambridge University Press.

Vygotsky, L. S. 1978. *Mind in society: The development of higher psychological processes*, edited by M. Cole, V. John-Steiner, S. Scribner, and E. Souberman. Cambridge, Mass.: Harvard University Press.

Wertsch, J. V. 1981. The concept of activity in Soviet psychology: An introduction. In *The concept of activity in Soviet psychology*, edited by J. V. Wertsch. Armonk, N.Y.: M. E. Sharpe.

◆ LINNDA R. CAPORAEL ◆

Discussion

Women are usually outsiders to war talk. Men fight; men control the institutions that declare and adjudicate war; men justify, explain, rationalize, and study war—and women look on. If we turn the tables and let women do the war talking, odd ideas emerge. Many of our "ordinary" understandings about war—understandings that we can view largely as men's understanding—seem puzzling and peculiar. For example, we expected primatologists to return from the field with data to answer the age-old question: are humans innately aggressive? Yet from this other side of the table, the expectation seems absurd. Half (in actuality, more than half) of the human species typically does not engage in war, and no one financed research to find out why. It is as if men saw themselves more closely related to nonhuman primates than to the female members of their own species. Not surprisingly, the primate data fueled a new round of evolutionary accounts that "explained" and justified aggression (e.g., Ghiglieri 1987; Wrangham 1986).

Are humans innately aggressive? The question, as Mary Brown Parlee points out, is scientifically useless. At best, it provides a benchmark that women can use to demand explanations for the gender difference. Either we are forced to develop two psychological theories, one for males and one for females (which is the usual strategy in the folk psychology of everyday language), or we have a single theory that requires us to examine both the everyday language of aggression as well as the differences in the situations men and women confront.

Aggression, Parlee notes, involves "at least an intention to hurt another person and actions to carry out the intention. . . . [A]ggressive acts are in part defined in terms of the opportunities afforded by particular settings." Different settings provide different opportunities for aggression, and different people are likely to respond to those opportunities differently. The correlated variations among settings, opportunities, and people, are the basis for appropriate scientific questions.

But, as the example from primatology suggests, the apparent correlation between combat and males is so pervasive that underlying cultural assumptions can replace empirical facts. It does make more sense to study related primate species rather than human females if women's "natural sentiment for peace" keeps them out of the fray. But such a sentiment is less an empirically demonstrated fact than an underlying assumption about the irrelevancy of women in discussions of war. One way to break such conceptual linkages that depend on what seems to be obviously "natural" is to invert points of view and categories—including those about sex and gender. Thus, we might ask if there are conditions when women engage in combat (ignoring the common wisdom that this is a silly question).

Women *do* fight. When war is literally at the threshold, sex disappears as a salient criterion for who may and may not bear arms in defense of home. Women have fought in the absence of both training and weaponry for combat. There also seem to be circumstances when women engage in group conflict specifically as women; such occasions may be interpretable as responses to attacks on gender roles. Although feminine identity is rarely threatened, such threats can occasion loosely organized violence by women—riots, rebellions, or "women's wars." The Igbo women's war (Sanday 1981), the babski bunty rebellions (Conquest 1986), the *femmes sans-culottes* of the French Revolution (Levy and Applewhite 1980), and American women's protests in the 1950s against strontium 90 in milk share a common defense of resources that are necessary for women's perceived roles in food production and procurement. Ardener reports that *anlu* (women's war) was a traditional response in parts of Africa to sexual insults—such as unflattering remarks about a woman's genitals—that devalued feminine social identity (1973). In 1958 seven thousand Kom women responded to an *anlu* call, not for sexual insult, however, but to protest colonial policies that reduced women's traditional control over farming and farming techniques. A number of women were killed or injured when the authorities shot into the crowd. Regardless of sex, it appears that people are sometimes willing to take considerable risks to preserve culturally constructed notions of gender roles.

Yet, surely war also interferes with women's performance of their perceived traditional roles: why don't women rise en masse to protest war? Judith Wishnia reports that women peace activists traditionally have been a small group. Most of them were feminists, but most feminists were not peace activists. The observation seems to belie the common perception that women are innately pacifistic or socialized to peace. Why, then, is there the widespread perception that women oppose war when so few of them act for peace during war? Women are de facto pacifists, less by choice than by laws that keep them from picking up arms in defense of their country. The connection between feminism and pacifism does not lie in their socially or biologically determined sentiments for peace, but in a *self-aware* political consciousness unifying peace and human liberation. The significance of such deliberate political choices can be appreciated against the psychological background of social identity.

There are many levels at which people identify—for example, individual, gender, ethnic, and cultural. Social identity takes the form of cognitive groupings of oneself and others, forming an in-group, along a dimension of contrast or opposition to some other group. One's own group—hence, social identity—is evaluated more positively to the extent that it is perceived as prototypical of the next more inclusive category (Turner 1987). I have suggested elsewhere that the cultural symbols associated with warfare—for example, the victory parade—have different consequences for men's and women's social identities (Caporael 1987). Women perceive these symbols as making a category distinction between their cultural group and the enemy group. Men perceive these symbols as making a category distinction not only between cultural groups, but also between males and females. For females, the parade is about "what our country did"; for males, it is about "what we men did for our country." For women, loyalty to country corresponds to loyalty to their men. In general, it takes little political deliberation to roll bandages, organize letter-writing campaigns and staff social clubs for "our boys." Such behavior can be an unsurprising result of more or less automatic "us and them" categories of social life. In contrast, it seems not only that women peace activists deviated from gendered and nationalistic expectations, but that they did it under circumstances when ingroup identification may be most compelling—when the group is at war.

Wishnia's paper, which confronts us with the words of women, also makes me wonder about their relationships with "their" men. Were the "peace feminists" resistant to nationalistic appeals because they had no men in war, or was it just the opposite? Did the peace feminists have husbands, brothers, or sons in battle? Did these women write their men, promising to bring their loved ones home by their peace efforts?

Is there a systematic relationship between women's efforts for peace and their connection to men at war?

I am not certain how Sally L. Kitch might deal with the last question. In her article she describes little connection between males and females. It is a mistake to view a "we-they" understanding of the world as an attribute of the male gender ethos; the Samantha vignette Kitch analyzes so well is underwritten by we (women) versus them (men), as is Kitch's essay (a delicate issue, to which I will return later).[1] We-they categorizations may be an endemic feature of human thought (Caporael, Dawes, Orbell, and van de Kragt, 1989).

Yet, Kitch is profoundly correct that cultural concepts of masculinity are tied to warfare. Sanday (1981) and Turney-High (1971), among others, report that males in premodern groups claim they "are no longer men" because European intervention ended traditional war activity. The association is also explicitly identified and defended in the contemporary American military (Golightly 1987). When women do take the role of warrior, it appears that they are required to reject functionally female and socially feminine identities by vowing to remain virginal and adopting the attributes of masculine identity. Joan of Arc is probably the most familiar example. Among certain Germanic groups, a woman could reject a potential husband chosen by her family by swearing virginity and adopting the clothing and life of a warrior alongside male warriors. The Dahomeans, who had a state-level social organization, had a corps of women warriors—all sworn (albeit imperfectly) to remain virgins until receiving the king's permission to marry.[2] These examples suggest Kitch's developmental analysis of difference in gender norms might be incomplete. There seems to be no essential, or biological, reason that women cannot become warriors. The reasons are social: the condition of warrior corresponds to masculine gender roles, not the male sex. Nevertheless, even if biology fails to explain this correspondence, there must be universal conditions of human existence that result in the repeated social constructions linking masculinity and war.

Unlike the condition of warrior, the condition of mother corresponds to the female sex, not to feminine gender roles. The act of giving birth creates an insuperable boundary between female and male in terms of what women can do and men cannot: women *can* be warriors (and in a pinch, they will fight), but is there anything as vivid, salient and dramatic as birth that men can do and that women cannot? Masculine identity is a problematic task for a culture, an issue that Kitch alludes to at the end of her paper. There is no "natural," in the sense of biological, contrast that correspondingly—for all time and for all audiences—defines the consequences of the category "male" for behavior as childbearing does for "female" (Stiehm 1981). There is no corresponding difference that can form a basis for a positively valued social identity. "Not childbearing" is hardly a behavior unique to men (or until recently positively valued)—the aged, the young, virgins, males, and women much of the time are "not childbearing." One common route to solving the problem of "what men do" is to culturally construct masculine social identity by opposing and contrasting it to an equally culturally constructed feminine identity consisting of prohibitions on and prescriptions for women.

Given the dramatic nature of birth, we might expect that it would be pivotal for the construction of gender by invoking symbolic similarities and contrasts. For example, among the Spartans, only two categories of people had their personal names inscribed in stone at their deaths: men who died in battle and women who died in childbirth. Spartan women enjoyed a high degree of social status and respect within their culture because women had a critical role in producing soldiers for the state. In Sparta, both sexes were perceived to endure pain and risk death for the same collective purpose. Similarly, the Ashanti also had rituals for women who died in childbirth that mirrored those for men who lost battles. Among the North American Papagoes, the construction of meaning takes the opposite direction, that is, from feminine to masculine. The scalp of an enemy is a "child" that brings satisfaction to the "mother"—the warrior who killed another man. In other tribal cultures, the connection between childbirth and war is also posed as a polarity between "making life" and "taking life," which nevertheless is linked because "blood binds the warrior and the child-bearing woman" (Sanday 1981, 44; see also Turney-High 1971).

So we should add to Kitch's developmental analysis the point of view of the child, who must construct his or her identity constrained by the structural conditions Kitch describes. Boys and girls are faced with different problems. Boys recognize that they are not like their mothers or sisters. Males face a category problem: how to classify two different entities, themselves and their mothers and female relatives, into a single category, human (cf. Ardener 1973)? When men are not involved with child rearing (remember Parlee's point about settings), boys have little opportunity to see other men engaged in nurturant activities. If masculinity is defined as that which is "not-feminine," boys have little experience about that which is masculine. They are left with a simple rule for constructing their own masculinity—the opposite of what females do, along with whatever characteristics of masculinity that the mother reinforces in her child rearing.[3]

This brings me to an issue I mentioned earlier. Kitch presents us with an anomaly; on the one hand, "ordinary women" find war disgusting; on the other, women control child rearing. Are we to believe that the mores of a misogynistic society are so powerful that they contravene the values and expectations a mother transmits to her children? That situation would not only devalue women's power, but it would also blind us to the possibility that women might play parts in the social construction of masculinity—and, hence, the maintenance of the war ethos. By adopting a we-they framework that presents women in positive terms and men in negative terms, Kitch invites us to be blind to the *human* dimensions encompassed by warfare. Yet it is precisely these dimensions that a feminist critique should illuminate, even if it means portraying women in less than glowing terms.

In focusing on what men do to make war and how men should change, Kitch's analysis inevitably becomes male-focused. It is the same focus that characterizes the thousands of volumes and papers written about warfare (cf. Caporael 1987; Stiehm 1984). In a paper on the Yanomamö Indians, for example, Chagnon argues that men go to war because it enhances their reproductive success; men who killed had more wives and offspring than men who did not (1988). Chagnon ignores the obverse. Women have reproductive interests that are arguably more compelling than men's, and they also have economic interests invested in males. Thus, we should expect women to strongly oppose war. What we find in Chagnon's paper, however, are illustrations of women playing active parts in the Yanomamö war ethos.[4] Yet these observations are not integrated into Chagnon's theoretical framework.

Similarly, women's roles in facilitating aggression are not addressed in Kitch's theory. Kitch allows women only a maternal power of restraint and forgiveness. Hence, the sole recommendation she can make is that politicians learn a female gender ethos. This analysis ignores the fact that political culture, as most aspects of culture, is superimposed on a sexual politic between males and females.[5] Remedies that focus only on the minority of men who hold political power will be minimally effective.

The problem is that the division of males and females into two different, unconnected groups does not allow for women to change men. Yet women participate in selecting and validating desirable attributes for masculinity in a culture. If ideals of masculine dignity rest on its "nonfemaleness," will women want "nonmasculine" men?[6]

Whether or not women can afford to select new masculine ideals may depend on their economic circumstances. (In what follows, I will be speaking in gross generalities, more with a view to inspiring more "woman-focused" thinking than to persuading the reader that the construction I offer is right.) Let us make a distinction between women with many choices and women with few choices. The range of choices a woman has depends on her social class background, education, employability, and so forth. The fewer choices a woman has to improve her economic circumstances—marriage being the most available route—the more likely it is that she will select a traditional man, who might subscribe to the war ethos but might also subscribe to the "good provider" ideal as part of the masculine ethos. For this woman control of sexuality, pregnancy, and child rearing are crucial for equalizing (to the extent it can be equalized) the power differential between male and female.

The greater scope a woman has to support herself and her children, the more she can afford to choose a man who deviates from traditional masculine ethos—or choose to have no constant male partner at all. In fact, we could go so far as to say women with many choices are able to shape new ideals of masculinity. A woman might have a child with minimal help from a male, but a male is unable to have children without considerable cooperation from a woman. Women with many options may be willing to relinquish some control of child rearing to men in exchange for greater female autonomy and greater male androgyny. These "new" masculine attributes could promise spiraling changes in gender roles because—to go back to our earlier points about child development—now boys and girls will be exposed to fathers with substantial child care responsibilities.

The framework proposed here differs from Kitch's in only one respect: it claims that we-they cognition characterizes both males and females, not just males. The *outcomes* of the two approaches are the same, but for different reasons. We want to see more women in political office, but not because women will bring a special touch to counter nationalism. Rather, the continued expansion of women's opportunities will bring men closer to home as they juggle the demands of child rearing, family life, and career with their partners. One agenda then has little to do with urging men to change; men will change when women prefer men who are different.

NOTES

1. The *Lysistrata* by Aristophanes, which portrays the women of Greece undertaking a sex strike to end war, is frequently touted as a "peace play." It is not. The women object to the men fighting other Greeks when they should be fighting non-Greek barbarians. The play ends with a celebration of the honors and glories of past wars and with ritual praise to Athena, "the goddes unvanquished in war."

2. Such vows are probably anachronistic in modern states, although the expectation to adopt masculine identity might express itself more subtly. When

Soviet women fighter-bombers began their missions under the leadership of a woman, she tolerated small, hidden infractions of regulations related to feminine identity (for example, long hair hidden under a cap and bows on underclothes [Myles 1981]). After her death on a combat mission and the assumption of the command by a man, such practices were prohibited vigorously.

3. Masculinity on these terms is a fragile thing, reflected in our language. There are no corresponding terms of sexual mutilation or incompetency such as "emasculating," "castrating bitch," and "impotent" that derive from femininity. Where the capacity for childbearing provides a concrete basis for a woman to construe a positive social identity, a man might wake up each morning wondering what he will have to do today to prove that he is "not a wimp."

4. For example, "the headman of one of the smaller villages . . . was killed by raiders in retaliation for an earlier killing. His ashes were carefully stored in several tiny gourds, small quantities being consumed by the women of the village on the eve of each revenge raid against the village that killed him. According to the Yanomamö, women alone drink the ashes of the slain to [fill the raiders] with resolve. In 1975, 10 years after his death, several gourds of his ashes remained, and the villagers were still raiding the group that killed him, who by then lived nearly 4 days' walk away" (Chagnon 239). According to Chagnon, consuming the ashes, which only the women do, puts the men into a frenzy (989). Women's influence is felt informally as well as ritually. One man led a raid on a distant village that had done neither him nor his kin harm, because his wife grieved so much for her sister, who had been killed by members of the village. The man "reported that his wife was very pleased with his actions" (991, n. 25). Although the Yanomamö are a male-dominated culture, Yanomamö women are not powerless. Rape is rare; women allow themselves to be openly seduced by men from neighboring groups, and, if a man cannot support his wife, she may take a second husband (992, n. 42). In just what sense are we to understand Chagnon's—and the Yanomamö's—claim that their wars always start over women?

5. This is the lesson in an old (probably familiar) joke. An average sort of man has been asked how he and his wife make decisions. He explains that he makes all the big decisions, such as whether or not the United States should support a nuclear reduction in NATO countries. He lets his wife make all the little decisions, like whether they should buy a car or a refrigerator this year and where the kids should go to school. The story subverts academic notions of dominance by locating the real source of social power in individual lives in the household rather than in the public domain. Conscious of this politic (despite simultaneously glorifying war), English critic and reformer, John Ruskin (1866–1962) recommended that every upper-class European woman vow to wear black—without ornaments, jewels, or excuses for "evasions into prettiness." He predicted no war could last for a week.

6. For an elaboration of this idea, see Anaïs Nin, *"In Favor of the Sensitive Man" and Other Essays* (New York: Harcourt Brace Jovanovich, 1976).

REFERENCES

Ardener, S. G. 1973. Sexual insult and female militancy. *Man* 8:422–40.

Caporael, L. R. 1987, November. A window on war: Women and militarism in classical Greece. Paper presented at the meeting of the American Anthropological Association, Chicago.

Caporael, L. R., R. M. Dawes, J. M. Orbell, and A. J. C. van de Kragt. 1989. Selfishness examined: Cooperation in the absence of egoistic incentives. *Brain and Behavioral Science* 12:638–739.

Chagnon, N. 1988. Life histories, blood revenge, and warfare in a tribal population. *Science* 239:985–92.

Conquest, R. 1986. *The harvest of sorrow: Soviet collectivization and the famine-terror*. New York: Oxford University Press.

Ghiglieri, M. P. 1987. Sociobiology of the great apes and the hominid ancestor. *Journal of Human Evolution* 16:319–57.

Gilligan, C. 1982. *In a different voice*. Cambridge, Mass.: Harvard University Press.

Golightly, N. L. 1987. No right to fight. *Proceedings of the U.S. Naval Institute* (December):46–49.

Levy, D. G., and H. B. Applewhite. 1980. Women of the popular classes in revolutionary Paris, 1789–1795. In *Women, war, and revolution*, edited by C. R. Berkin and C. M. Lovett. New York: Holmes and Meier.

Myles, B. 1981. *Night witches: The untold story of Soviet women in combat*. Novato, Calif.: Presidio.

Ruskin, J. [1866] 1964. *The crown of wild olive*. Reprinted as War. In *Man and warfare*, edited by W. F. Irmscher. Boston: Little, Brown.

Sanday, P. R. 1981. *Female power and male dominance: On the origins of sexual inequality*. New York: Cambridge University Press.

Stiehm, J. H. 1981. *Bring me men and women: Mandated change at the U.S. Air Force Academy*. Berkeley: University of California Press.

Stiehm, J. H. 1984. The man question. In *Women's view of the political world of men*, edited by J. H. Stiehm. Ardsley-on-Hudson, N.Y.: Transnational.

Turner, J. C. 1987. *Rediscovering the group: A theory of self-categorization*. New York: Basil Blackwell.

Turney-High, H. H. 1971. *Primitive war: Its practice and concepts*. Columbia: University of South Carolina Press.

Wrangham, R. W. 1986. The significance of African apes for reconstructing human social evolution. In *The evolution of human behavior: Primate models*, edited by W. G. Kinzey. Albany: State University of New York Press.

♦ III ♦

Who Fights for Peace and Who Makes War?

♦ REGINA E. WILLIAMS ♦

Jahtime

"Don't be afraid of
atomic energy 'cause none
of them can stop
Jahtime."

—"Redemption Song"
Robert Nesta Marley

My brother
you seek revolution
in a rhetoric
of stone and bullet
you pit beard against beard
(Is it Che or a high Holy Man
squatting in the reeds?)
you separate camouflage attire from saffron robe
Tell me, is one more powerful or valid
than the other?
you pine for the rhythm
and fervent cadence of a Lucy Parsons
Frederick Douglass or Malcolm
forgetting that for each
there are thousands unnamed and faceless
flexing their muscles against the same oppressor

There are no masses, my friend
only you and I
pitching stones, swinging machetes
and uttering our silent prayers
there are no masses

just you and I
separate wings of the same soaring eagle
no masses
only sojourners seeking truth

My brother,
you think a woman with a hoe
or rock or skillet,
quilting her private patterns
less essential than a gun-toting soldier?
you think the cant of a beggar priest
less powerful
than the thunder of Gandhi?
Are they not the same
winged dream exalted?

Tell me Maggie Lewis didn't pay the price
Tell me Lahaima died in vain
Tell me Gladys V. Thorne was not a revolutionary
Tell me we don't march to the same war

The symbols and signs are all here
my friend:
the catclaws highlighting my eyes
the white in my lashes
the blue ring orbing my irises
they are here my friend
all you need do is listen
see, continue the journey
and believe

Each time the fiery and swirling sun
splinters and splits the deep dark horizon
there is a revolution
Each babe born
breeds revolution
Each blade of grass
that cracks the contour of a rock
Each blistering day
there is revolution
my brother

Retrain your ear for the rumbling sound
sharpen your sense of smell
taste the sweet nectar of movement
tough the fiber of each thread woven

there is revolution all around you
brother

There are no masses
just you and I
sucking on sugar-teats
and humming the history
that leads us forward
only you and I
a measure of Jah's time
for this time
brother

♦ DAISY GARTH ♦

The Present Situation of Nicaraguan Women

The reflections presented here are intended to give an idea of the present situation of women in Nicaragua—a society that in this stage of the "Revolucion Popular Sandinista" is in political, social, and economic transition at the same time that it is the victim of constant imperialist aggression. The historic dependent position of and discrimination against Nicaraguan women is a fundamental element in our ever-intensifying struggle for the triumph of our demands. Therefore, to understand these demands, whose successful resolution will form a part of the future transformations of the new society Nicaraguans are constructing, it is necessary to consider the history of women's oppression in Nicaragua and the measures taken in favor of women since the triumph of the revolution in 1979. These measures are a manifestation of women's increasing integration into the political, economic, and social life of the country. This essay will examine the specific conditions in which new policies have developed in Nicaragua, the daily conflicts created by them, and women's responses to these conflicts.

HISTORIC ANTECEDENTS

Imperialist domination, dependent capitalism, and the Somoza dictatorship are the fundamental elements that for decades exposed the

Nicaraguan people to the most cruel exploitation, oppression, and discrimination. We Nicaraguans were denied the right to the most elemental necessities of life such as good health care, decent education, proper nutrition, and adequate housing. Nicaraguan women were immersed in this context as victims of these and other forms of oppression, specifically those derived from our position as women in a "machista" society. In addition to our struggle against the historic traditions within Nicaraguan society that have placed us in a secondary position, as women, we must confront other, more powerful enemies, such as the transnational feminine model created by the capitalist world, in order to guarantee our survival.

Statistics show that, in 1970, 48 percent of the heads of families in Managua, the capital of Nicaragua, were women. In rural areas of the country this percentage increased to 53 percent. On the other hand, in 1975 women represented only 39 percent of the economically active population of Managua. Of those employed, the majority worked outside the formal sector of the economy; 70 percent were in personal services, and 55 percent were street peddlers, small merchants, and day workers (Guido 1975). Although women represented a high percentage of the principle supporters of families, the figures demonstrate their comparatively low participation in the labor force and especially in the higher-paying jobs of the formal sector.

In the prerevolutionary period few women held high positions in state institutional structures. Those who did encountered problems such as unequal pay between the sexes for similar work. Women were considered to be less intellectually capable than men. If a woman were an accountant undertaking the same work as a male accountant, for example, the latter would always receive a better salary. Women were so accustomed to this state of affairs that most would not even consider applying for a job at a high level. Instead, they solicited work in middle- or low-level positions. In many instances, job advancement or even retention depended on a woman's willingness to accede to the sexual advances of her male superiors in the workplace.

Pregnant women were unlikely to obtain employment at all. In fact, employers felt themselves to be within their rights to fire recently employed women who turned out to have been pregnant at the time of their initial hiring. Women who became pregnant on the job were given maternity leave and found it very difficult to return to their jobs.

In general, women were excluded from work outside the home. The work of women was viewed generally as something complementary, not fundamental, and women themselves viewed it as such. They were supposed to dedicate themselves only to housework and raising children—a form of labor that was not considered productive work because, supposedly, it had no economic value. There were no retirement or social security funds associated with it.

The deformed system in which women were forced to conform to the unjust dictates of society contributed little by little to the creation of a new consciousness among women. They began to understand that there had to be basic transformations in the political and economic structures of society. In this way the women's struggle merged with the armed struggle of all the Nicaraguan people to defeat the Somoza regime and institute a revolutionary state.

To accomplish the change many women participated as couriers, providers of security houses (dwellings from which the Sandinistas operated), student activists, clandestine fighters, and *guerrilleras* in the mountains. During this difficult armed struggle, many women were harassed, jailed, tortured, raped, and assassinated by the National Guard. The struggle continued with the active and crucial participation of women and culminated in the triumph of the Revolucion Popular Sandinista on July 19, 1979.

THE PRESENT SITUATION

In the late 1960s, the FSLN (Sandinista Front for National Liberation) contemplated in its historic program the emancipation of women. In 1969 the first women's organization directed by the FSLN was formed. This group, the Alianza Patriotica de Mujeres Nicaraguenses (Patriotic Alliance of Nicaraguan Women), was made up of previously isolated groups of women who acted as advocates of political prisoners who had participated clandestinely in the ranks of the FSLN.

Later in 1977 the group was transformed into the Association of Women Confronting the National Problem (AMPRONAC) (Envio 1987). AMPRONAC began to bring together and analyze women's concerns as a starting place for the work of emancipating women. It was transformed again into the Women's Association: Luisa Amanda Espinoza (AMLAE)—a movement with an explicitly revolutionary character, named in memory of the first woman member of the FSLN to die in combat. (Espinoza was assassinated by the National Guard in Leon in 1970.)

As a woman's organization, AMLAE has made great strides. The post-triumph activities of AMLAE and those of women in general have been shaped by the varying realities of the revolution in Nicaragua. The process of consolidation of the women's movement and our general empowerment during this period has undergone four distinct phases. For clarity, I will discuss the women's movement within the context of each.

In the first phase, from 1979 through the middle of 1981, AMLAE took as its task participation in the reconstruction of the country. Members of this group and women in general participated actively in social projects. The literacy campaign helped thousands of Nicaraguans to improve their lives. Women worked as volunteers, teaching reading and writing. Many followed up these activities by participating in the program for adult education in which reading and writing skills were taught to those who were unable to obtain these skills through the literacy campaign.

Women also participated actively in social projects such as the health campaigns. Here they were active as health educators in the neighborhoods, bringing health care to children and vaccinating them in the popular campaigns, thus preventing serious illnesses. They also collaborated in campaigns for better hygiene and helped to clean neighborhoods in order to reduce the rate of infant mortality. They were concerned about and educated around the topic of health problems that confronted women in the communities because of the lack of doctors, health centers, and adequate treatment.

Women were heavily involved in the establishment and operation of the revolution's mass organizations, such as the Committees for Sandinista Defense (CDS), the *expendios populares* (popular stores that provide people with basic foods in order to guarantee an equitable distribution of food), the Juventud Sandinista 19 de Julio (a revolutionary youth organization), the various party organizations of the FSLN, the National Union of Farmers and Cattlemen (UNAG), and a series of political and cultural activities. The prominence of women in these activities is important to emphasize because for the first time, while still playing their traditional roles, women became involved in the political life of the country and emerged as visibly present and influential forces in Nicaraguan society.

We can say that this first phase signified the beginning of the new conception of women's roles in Nicaragua. We were now seen as persons who can participate harmoniously in all aspects of Nicaraguan life—political, economic and domestic.

The second phase in the process of the empowerment of Nicaraguan women occured within the context of increased military mobilization in response to a growing counterrevolutionary war and lasted from 1982 through 1984. The "contra" war imposed by the United States required the total dedication of the country to its defense and created specific conditions to which women had to respond. Just as in the past, when they had engaged in the struggle to defeat Somoza, women actively participated in the defense of our country during this period. Many women requested to be trained militarily and entered the Militras (voluntary local military units).

With so many people involved in military activities, an acute labor shortage developed in crucial areas of the Nicaraguan economy. As more and more men were called into the military service, even more women assumed the role of heads of their families. At the same time the Nicaraguan economy entered a period of crisis—a result of the North American economic blockade, worsening warfare, and a more widespread international economic crisis. In the everyday struggle to support their families within this context, many more women than before were forced into what could be termed a double or even triple workday.

Most women have need to take on some kind of job outside the home to supplement their families' incomes, not simply because they enjoy the work but because poverty forces them to assume double burdens. At the end of the workday spent outside the home, most women must return home to face a variety of domestic chores. Domestic labor in Nicaragua—as in most underdeveloped countries—is much more demanding than in the developed world. For example, women must wash clothes by hand and prepare meals, most times using charcoal or firewood. Then, because of acute shortages of food and other essential items, they must go to a variety of locations, sometimes far from their homes, and get in line to obtain the items necessary for themselves and their families. The situation has been and still is extremely difficult.

Women entering the job market have been faced with a limited number of options. Because of the distorted nature of the war-ravaged Nicaraguan economy, some women have opted to participate in the informal sector of the economy. They have become sellers in the markets speculating in products and prices. Many are squatters living in shacks in vacant lots within urban centers. AMLAE has been actively trying to integrate this group of women into the productive activity of the country.

On the other hand, women have been breaking down the old barriers that once kept them from participating as equals in the formal sector of the economy. The rapid increase of men joining the military in the rural areas has been a fundamental factor in women's increasing participation in agriculture. Little by little, they have ceased to be a temporary work force (as they have been, historically) and are now becoming a permanent work force that guarantees production.

Women have joined the cooperatives and the Association of Farm Workers (ATC) in their family production units. Within these organizations, they work side by side with and equal to men, cooperating in the betterment and growth of their work units and this form of organizing work. This development demonstrates that women are willing to undertake all of the tasks that contribute to the reconception of their traditional roles.

During this phase women also became much better represented as workers in the food, textile, and other industries. Generally, in these activities women still face many problems. They continue to be placed in jobs reserved for women; increasingly, however, they are attaining positions in maintaining and servicing machines, transporting materials, and operating machinery, which are considered to be traditionally masculine jobs. Women are also now union members. They have been able to take their demands to representative organizations and propose possible solutions to problems.

The increased presence of women in the work force necessitated the creation of child care centers (CDIs) nationwide. CDIs have gone a long way toward responding to the problems that confront working women in the care of their children.

The heightened economic role of women was initiated in this second phase and was consolidated during the third phase of the movement towards women's liberation, a period that began in 1984 and extended through 1986. In the midst of the problems and achievements of women and the society as a whole during this period of the revolution, women's rights continued to be discussed on the national level. The deliberations in the National Assembly during this period are perhaps most representative of these activities. First and foremost, the women's movement was granted formal representation in the assembly.

In this forum the paternity law was discussed. The aim was to break with the traditional style of paternal jurisdiction (*patria potestad*). The law, as previously promulgated, guaranteed legal recognition and rights of support only to children born in wedlock. Under this law there were many cases of men with numerous children by a number of different women in which the responsibility to raise and support the children fell solely on the women, who had no legal recourse because they were not legally married to the children's fathers. There were other cases of women with nine or ten children whose father took no responsibility to feed them and were not legally required to since no wedding had taken place. During the period from 1984 through 1986, this law was abolished, and the law between "Madre, Padre e Hijos" (mother, father, and children) was enacted, making both parents responsible for their children regardless of their marital status. Now the essential element of the law is not legal marriage but responsible paternity. We are attempting to establish the conditions necessary to create a consciousness of responsibility for children.

The establishment of the law of Equal Salary for Equal Work has also been achieved. The law has had the biggest impact in rural areas where many women and children work in farming activities, receiving little or no salary. The discrepancy occurred for two reasons. First, because their labor was considered part of a family salary, it went to the male head of the household. Second, this type of work was considered traditionally to be men's work; thus, women were thought not to be as capable of performing it and were paid less accordingly.

The assembly has also considered the issue of abortion, although no decision has been reached. Abortion is a very delicate issue because of its religious implications. Unfortunately, while abortion remains illegal in our society, it is performed daily in unsafe and often unsanitary conditions and is a tremendous health hazard to Nicaraguan women. What is more, because of religiously based aversion to any public discussion of pregnancy and abortion, young women in general do not even have access to the type of information that would be useful to them in avoiding unplanned pregnancy.

Changing the archaic divorce law inherited from the previous regime has also been discussed in the assembly. If the new law proposed by AMLAE is approved, unilateral divorce will be possible for the first time.

Besides their work in the assembly during this period, AMLAE has been able to achieve a number of other advances. Historically, there have been many cases of women who have suffered physical mistreatment and disrespect to their persons, which they have had to tolerate. AMLAE has collected a series of opinions from the different sectors of Nicaraguan society in an attempt to bring together the ideas of each and find a solution to this problem. This type of abuse has been condemned in the legislature. To deal with cases of abuse and other legal problems facing women, AMLAE has created the Woman's Legal Office, which handles serious cases, such as single mothers unable to get fathers to support their children, women's physical mistreatment by men, divorces, and a series of other problems specific to women. The office also performs the important task of advising women, following up on their cases, and referring them to the appropriate state institutions such as the courts, the office of family protection at the National Institute of Social Security and Welfare (INSSBI), and the police, when the situation calls for it.

AMLAE has also created programs in mass communications where for the first time the oppressed position of women is being publicized. These programs also seek to clarify to Nicaraguans the social transformations that have taken place concerning women. They also aim to create consciousness among men seeking their support so that, united, we many combat inequality and disrespect.

Even though machismo has been the object of much discussion in the assembly as well as in the society at large, this phenomenon is the heritage of the past and will take a long time to overcome. Despite our advances, women continue to be confronted with a difficult situation. Even within the party structures of the FSLN, there are cases of sexist men. Many times revolutionary women are not accorded the respect they are due. As a matter of course, they are often expected to work outside the home, participate actively in the party, and then be responsible for all housework.

The fourth phase of the maturation of the women's movement extended from 1986 to February 25, 1990. It was characterized by the continued struggle for women's liberation, which continued to be waged within the specific conditions created by the war of aggression and the economic crisis. The crisis worsened to the point where the Nicaraguan economy was one of little more than survival. As women, we had to adjust ourselves to this reality. Within its context we continued to discuss our problems while promoting the integration of women into agricultural and industrial production and defense of the country.

The military situation improved toward the end of this period, and the revolutionary transformations of Nicaraguan society aimed at liberating all Nicaraguans became institutionalized. As a consequence, the fourth phase was characterized by increasing independence of the women's movement. There was seemingly no longer as great a need for the women's movement to be preoccupied with the general problems of the revolution, and there was more space for it to deal with those problems that specifically affected women. At this point, AMLAE and the FSLN decided that AMLAE would be much less tied to the needs and directives of the FSLN and, it was able to respond more agilely to the needs of women in all sectors of Nicaraguan society.

CONCLUSION

The struggle of Nicaraguan women to transform the inherited conditions of exploitation, oppression, and discrimination has been continually championed by the women's movement since the days of the revolution. During that period the FSLN and the women's movement worked in close cooperation in the creation of specific laws that guaranteed women's rights. They are the most progressive in the hemisphere.

The February, 1990, general elections brought a political change to revolutionary Nicaragua. The Sandinista Front was voted out of power. Those of us concerned with gender issues speculate that there will be a return to a situation similar to the prerevolutionary scene presented above. It is still premature to say. However, there has already been an attempt by the new government to regain control over women's issues. After the elections, the Union Nacional Opositora (UNO) government took over the National Women's Institute, the only women's research organization in the country. The action took place because it is important to the government to change revolutionary women's ideology. The new government has also begun to introduce a different kind of television commercial than had been produced under the revolutionary government. In the new commercials women's bodies are used as an essential factor in getting messages across—much like the advertising broadcast in the United States. In this new process of change, one can see clearly the ruling elite's intentions to subvert the active role of women in Nicaraguan society and make us submissive again.

REFERENCES

Berezhnaya, N., and E. Blinove. 1987. *Mujeres en la lucha por la paz.* Moscow.

CIERA and ATC. 1987. *Jujer y agro-exportaciòn en Nicaragua.* Managua.

Envio. 1987. *Mujeres mas espacio y mas voz.* Managua.

Guido, L. 1987. Apuntes sobre la situaciòn de la mujer en Nicaragua. *Cuadernos de Sociologia* (May–December).

Mondragon, R., and C. Decker Molina. 1986. *Participaciòn popular en Nicaragua.* Mexico.

Montenegro, S. 1987. Las ideas sobre la mujer. *Cuadernos de Sociologia* (May–December).

Revista Editorial Vanguardia. 1987. *Women and the Sandinista Revolution.* Managua.

Revista Somos. 1983. *Oficina Legal de la Mujer* (June).

♦ MAMIE JACKSON ♦

Science for Profit versus Science for People

In an article on disarmament written by a member of the Women's International League for Peace and Freedom (WILPF) and published in the league's magazine, *Peace and Freedom*, the author, Kay Camp, proposes that putting more women in governments would bring peace to this world ("On Genes, Gender, and Geneva," *Peace and Freedom* 45 [March 1985]). She bases her arguments on some scientific studies summarized in *The Tangled Web* by Melvin Konner, a sociobiologist. The studies purport to show that men are more aggressive than women, and that behavioral differences are due to the differences in the genes of men and women. These genes trigger hormones that shape the brain and thus direct behavior. Because I disagree with the hormonal explanations for peace and war, as a member of WILPF, I am responding to these views.

Science is said to be the quest for knowledge. Unfortunately, scientific knowledge is sometimes used for destructive purposes. In their search for understanding of how the mind works, for example, scientists discovered that the mind can be manipulated. If this knowledge is used subsequently to control minds, then scientific discoveries are being used for unethical and inhuman purposes. It is important to search for a cure for cancer; if, however, as an accidental side effect of the experimentation, the scientist produces a substance that is destructive to people or the environment, the question must be raised: how much longer should the research continue? The scientist, and not the financial sponsors (corporations, foundations, and government), should have control over the findings of the experiment.

Another question that must be raised is whether scientific knowledge should be used for human annihilation by developing bigger and better weapons for war. The quest for scientific knowledge must be controlled and guided by scientists and the public. Presently, technological discoveries are moving at a fast pace. Without constant monitoring, the final result might well lead to the end of the world.

We do not always see that the corporate greed for profit forces scientific workers to experiment and test without concern for where the results will lead. The major corporations (many of them multinationals) separate the workers from each other so that knowledge cannot be easily shared, notwithstanding the scientific organizations. Only when a corporation believes it is safe does it allow the scientists to "publish" the results of their research. Because research is so costly, the individual scientist cannot afford the luxury of being "self-employed." S/he becomes, in the fullest sense of the word, a worker earning wages. As a worker, the scientist needs a labor union to—among other things—help maintain scientific integrity, monitor the use of scientific findings, and be able to eliminate any disastrous effects of the findings from public and private use.

With this advice to scientists as a background, it is time to approach the question of genes, gender, and war. It is not women or men who make wars. It is power, profits, and greed. Men became synonymous with war when the process of inheritance of wealth from father to son became the social custom. This gender role, historically assigned by society, is still practiced today. The notion that genes control gender roles can be compared to the idea that the sun, and not the earth, was the center of the universe. Genes do not program women to nurture and men to be breadwinners. Certainly, there are many examples today of women working and of child care being shared by husband and wife. These functions are therefore interchangeable.

The sociobiologists, described in *Peace and Freedom*, who tell us that genes determine gender roles and that gender determines who are the proponents of war, are putting out false propaganda. These scientists are allowing themselves to become the whip to keep us in place; they are intellectually attempting to brainwash us in order to remove the blame of war and destruction from the powers who profit from these wars. Men are not genetically aggressive, nor are women genetically passive. We simply have to look at the anti-choice supporters ("right-to-lifers") as a case in point. Women anti-choice supporters are just as aggressive as their male cohorts when it comes to saving the fetus. Similarly, women are as aggressive as men in sending their sons off to war. Just as women have aggressive potential, men have peacemaking potential. Movements for peace are, and always have been, made up of both women and men.

The Women's International League For Peace and Freedom is an organization that has always been in the forefront of the struggles for peace and women's rights. These struggles, however, have to be based on sound science. The reports of the sociobiologists that men are genetically more aggressive than women has been disproven in many recent studies by scientists who practice "good science" (see Part 1 of this volume). To stop wars, we must eliminate the profits of war, and to do that we need all the good women and men to participate in the struggle.

♦ ELLEN DORSCH, JOY LIVINGSTON, AND JOANNA RANKIN ♦

If Patriarchy Creates War, Can Feminism Create Peace?

Every day while we work, study, love, the colonels and generals who are planning our annihilation [make] 3–6 nuclear bombs. . . . They have accumulated over 30,000. . . . They have proclaimed Directive 59 which asks for "small nuclear wars, prolonged but limited." They are talking about a first strike. The Soviet Union works hard to keep up with United States initiatives. . . . We are right to be afraid. . . .

There is fear among the people, and that fear, created by the industrial militarists is used as an excuse to accelerate the arms race. "We will protect you . . .," they say, but we have never been so endangered, so close to the end of human time. . . .

We women are gathering because life on the precipice is intolerable.

We want to know what anger in these men, what fear which can only be satisfied by destructiveness, what coldness of heart and ambition drives their days.

We want to know because we do not want that dominance which is exploitative and murderous in international relations, and so dangerous

The authors are members of the Burlington, Vermont, chapter of the Women's International League for Peace and Freedom.

to women and children at home—we do not want that sickness transferred by the violent society through the fathers to the sons. . . .

We want to be free from violence in our streets and in our houses. . . . We want an end to the arms race. No more bombs. No more amazing inventions of death. . . .

We will not allow these violent games to continue. . . .

We know that there is a healthy sensible loving way to live and we intend to live that way in our neighborhoods and on our farms in these United States, and among our sisters and brothers in all the countries of the world.

—Unity Statement, Women's Pentagon Action

Few doubt the extremity of our current situation, that "we have never been so endangered, so close to the end of human time." We have even begun to understand the various military, social, economic, and moral dimensions of our situation, yet few of us act on our knowledge of these matters.

Feminists, in particular, have been slow to perceive militarism as a critical issue for women, although compelling evidence to the contrary has long been available. Many have not wished to associate themselves with the anti-feminist position that women want peace because we are "naturally" peaceful—the position so dear to the ideologues of the New Right and rationalized by the "scientific" principles of sociobiology, which are the subject of this conference.

We argue that militarism is a central concern for women in our struggle for liberation, but it is only a beginning. Feminism illuminates the reasons that our patriarchal culture has such difficulty in acting to avoid catastrophe. The patriarchal domination of women by men is intimately connected with the sweeping militarism that marches us ever closer to the brink. The essay begins with a critique of militarism from a feminist perspective, then connects it with the larger issue of patriarchal power as dominance. Finally, it examines how these considerations affect both our daily lives and our political work.

A FEMINIST ANALYSIS OF MILITARISM

War, we might say, is the organized pursuit of coercive power by overtly violent means. The military and its supporting institutions sustain the threat and actual practice of warfare. And in our time they pervade every aspect of our society and entail primary roles not only for armies and navies and for governments and industries—but also, as we shall see, moms and the "girls back home." Where military values shape civilian society or military people or considerations effectively control the state, we have militarism.

The essence of military institutions is coercion through violence. War and militarism have the primary function of imposing dominance over others—that is, of *winning*. Militaries both threaten and actually carry out physical violence, and for many purposes the threat is as effective as the actual commission of human slaughter. Thus, there is no sharp demarcation between threats and the actual use of military violence to the extent that, indeed, both impose dominance. In fact, for many purposes it is valid to consider that violence *is* the imposition of dominance (see Starhawk, for example).

Ellen Elster offers a "working definition" of patriarchy as "a set of beliefs and values supported by institutions and backed by the threat of violence. It lays down the supposedly "proper" relations between *men and women*, between *women and women* and between *men and men*" (1981). Barbara Zanotti suggests that "patriarchy is the rule of the fathers. It is a system of male headship, male domination, male power—a system of controlling woman through economic dependence, violence, and domestication—a system which assigns women to the private sphere of home and family and directs males to the public sphere of work and decision-making" (quoted in Reardon 1985). Both definitions identify patriarchy as a system of violent coercion. How then is militarism informed by patriarchy?

First, war and militarism foster and legitimize the prevailing system of patriarchal power, and they model and legitimate violence as an acceptable technique for achieving social objectives. Military training and values are seen as applicable and useful in "civilian" society at large. Indeed, military ability—which, presumably, means the ability to commit physical violence—is a major component of many males' gender identification. Masculinity is seen as primary to war and militarism; in fact, it has been *defined* by an ability to learn the necessary techniques of war. And because of its perceived importance, this ability is reproduced and enforced by a whole range of influences (parents, schools, and sports, for example, consciously or unconsciously, in successive generations of children.

Second, under patriarchy, women are coerced and encouraged to collaborate with the military on many levels, and women's wide participation is absolutely necessary for its continuance. It is an old truth that women have great power to limit and end men's wars, and throughout history groups of women have taken valiant steps to do so—Trojan women in Hellenic times, Iroquois women in 1590, and the Women's Peace Conference in The Hague during World War I, to name a few. But, more frequently, women have been drawn into participation in war as military wives, workers in military industries, mothers of sons, military service workers, military nurses, paid or gratis recreation workers for soldiers (such as prostitutes), a reserve work force to replace men in civilian work, and, occasionally, actual combatants.

In her 1983 book *Does Khaki Become You? The Militarization of Women's Lives*, Cynthia Enloe traces a number of women's military roles and outlines what surely remains *the* primary contradiction of militarism: war and the preparations for war are absolutely dependent upon women's efforts, and military establishments everywhere and at all times have been concerned fundamentally with denying this. Given that at least ten support people are now required to field every actual combat soldier, it is women who make the military possible, but still the quintessentially masculine activity of soldiering must not be sullied by an overt or apparent female presence.

The U.S. military, for instance, is concerned with women in many ways:

- Soldiers with "military wives" (this is a military term) are more docile and predictable, and their reenlistment is found to depend heavily on whether their military wives are contented.
- Extensive arrangements are made to facilitate and control prostitution in the vicinity of military bases, particularly outside the United States.
- Women constitute an essential source of cheap labor both directly and indirectly available to the military, and any sort of pay equity would make military bills much more expensive.
- The military is concerned with children—that is, male children—and the so-called family environment. A sufficient number of boys must be born and raised under conditions conducive to their becoming "good soldiers"—and to this end no one has invented a better such environment than the gender-polarized, patriarchal nuclear family.

Third, patriarchy is also connected to militarism in that the military exercise of coercive power necessarily involves the state. Coercive power is the universal prerogative of the nation-state through the law, or by extralegal means (such as the CIA or KGB), and this prerogative is backed up by civil police power and ultimately guaranteed by military force.

Indeed, the state and the military are so closely identified that we can hardly think of them separately. We can debate where power lies and prefer the U.S. system perhaps (where the military is sometimes partially controlled by the state) to the junta (where the military and the state are virtually synonymous), but nonetheless we have difficulty distinguishing between the state and the military. The close relationship between the military and the state is easily seen in most countries, but, in civilian governmental systems like our own, it is heavily obscured and mystified. The point here is that the state and the military are *not* distinct but are two faces of one system of patriarchal power.

In this context, it is then clear that militaries function much more regularly as guarantors of domestic political arrangements than as protectors of a people against external threats. That this is true in many instances is patently obvious, although, initially we may question whether it is true of our own country. Consider that in the United States the specter of external threats always raises strong appeals to bridge or defer internal divisions. Indeed, external "enemies" have often appeared just in time to divert the nation from pressing internal change. World War I, the Cold War, and Vietnam are all examples here. No matter what issues are raised, we are instructed to behave and be quiet until the threat is over.

Finally, feminist insights into the patriarchal use of violence in such instances as rape and battering illuminate our understanding of war and militarism. Rape and battering are understood as *political* acts, violent acts of terrorism in the service of coercion and domination. Indeed, according to Susan Brownmiller in *Against Our Will*, rape is "nothing more or less than a conscious process of intimidation by which *all men* keep *all women* in a state of fear."

War and rape are thus closely connected. Both *intend* coercion or domination through violence, and both can be group as well as individual acts. If rape is generalized to include other acts of similar intent—such as battering—then the analogy to war as rape of *others*, rather than women, becomes inescapable. It is hardly surprising that patriarchal civilization endlessly confuses sex and love with coercion and violence. The link is rape, and the "highest expression" is war.

Military misogyny is so intense and repulsive that we might be tempted to overlook it, regarding it as incidental to what is, after all, a nasty business. We must steel ourselves, however, to look more closely in order to understand the relation between war and rape. Militarists of all stripes, from the infantryman to the industrialist, cultivate misogyny (as well as racism and classism) because they must learn to dehumanize their potential victims. So woman hating through the pornographic objectification of women, for instance, provides essential training for soldiers whose ability to kill efficiently requires them to sever all human connection with their victims. The lesson is all too well illustrated by the boot camp training jingle that goes:

> This is my rifle (slaps rifle),
> This is my gun (slaps crotch),
> One is for killing, the other for fun. (Quoted in Warnock 1982, 22)

Other examples abound. In his book *Sexual Suicide*, George Gilder writes of Marine Corps boot camp:

> From the moment one arrives the drill instructors begin a torrent of misogynistic and anti-individualist abuse. The good things are manly and collective; the despicable are feminine and individual. Virtually every sentence, every description, every lesson embodies this sexual duality, and the female anatomy provides a rich field of metaphor for every degradation.
>
> When you want to create a solidarity group of male killers, that is what you do, you kill the woman in them. That is the lesson of the Marines. And it works. (Quoted in Daly 1978, 358)

In *A Rumor of War*, Phillip Caputo, who was himself a Marine infantry officer in Vietnam, writes:

> The tedium was occasionally relieved by a large scale search-and-destroy operation. . . . Weeks of bottled-up tensions would be released in a few minutes of orgiastic violence, men screaming and shouting obscenities above the explosions of grenades and the rapid, rippling bursts of automatic rifles. (Quoted in Daly 1978, 358)

Mary Daly writes in *Gyn/Ecology* that "the bonding of trained killers requires the perpetual semantic degradation of women, in an effort to kill male weakness, which is misnamed 'the woman' in them" (1978, 358). Thus, obscenities reflect the training of the drill instructors and are merely the rote, not the substance, of the lesson. The substance is the capacity to commit violence, and, when learned, rape and killing become virtually indistinguishable acts of war. The following is a description by a Marine sergeant of a gang rape he witnessed while in Vietnam:

> They were supposed to go after what they called a Viet Cong whore. They went to her village and instead of capturing her, they raped her—every man raped her. As a matter of fact, one man said to me later that it was the first time he had ever made love to a woman with his boot on. The man who led the platoon, or squad, was actually a private. The squad leader was a sergeant but he was a useless person and he let the private take over his squad. Later he took no part in the raid. It was against his morals. So instead of telling his squad not to do it, because they wouldn't listen to him anyway, the sergeant went into another village and just stared

> bleakly at the ground, feeling sorry for himself. But at any rate, they raped the girl, and then, the last man to make love to her, shot her in the head. (Quoted in *Bulletin of the Atomic Scientist*)

Misogyny is closely connected with the execution of military power at the highest political levels, as well. Lyndon Johnson was known to respect and value only tough, "real" men who were confident and hawkish about the Vietnam War. In *The Best and the Brightest*, David Halberstam writes: "Hearing that one member of his administration was becoming a dove on Vietnam, Johnson said, 'Hell, he has to squat to piss.' . . . Doubt itself, he thought, was almost a feminine quality, doubts were for women" (quoted in Daly 1978, 359). And in his biography of Richard Nixon, Bruce Mazlish describes him as follows: "He was afraid of being acted upon, of being inactive, of being soft, of being thought impotent, and of being dependent on anyone else" (quoted in Daly 1978, 359). What then are we to do with Reagan, and now Bush?

We should not think, however, that only crude soldiers and politicians connect misogyny and violence. Indeed, the connection can be intellectualized and abstracted so as to carry appeal to men far removed from the military. Among scientists and engineers, a distinction is made between "sexy" weapons and those that are not. Sexy weapons are "manly" and "fair." Sexy weapons are those of phallic form that are powerfully ejected by men and that kill, in turn, by forcibly entering a person or a city, preferably punctuating their entry with an explosion of orgiastic violence.

All manner of lethal military technologies can be suggested that are distinctly "unsexy" (such as electrocution, genetically engineered poison bees, or droves of little helicopters with nooses that seek out and snare their victims, hanging them until they are dead), but military minds universally prefer the sexy sort. The phallic imagery of most weapons—from guns to bullets to bombs—then is hardly imaginary or accidental; rather, it is a symptom of the disease. By way of example, a plaque hangs in the foyer of the Syracuse Research Corporation, a private "think tank" with large military contracts. The plaque is illustrated with a missile in flight, and its inscription reads:

> I LOVE YOU BECAUSE:
> —Your sensors glow in the dark
> —Your sidelobes swing in the breeze
> —Your hair looks like clutter
> —Your multipath quivers
> —Your reaction time is superb
> —Your missile has thrust; it accurately hones in on its target

—The fuse ignites, the warhead goes;
SWEET OBLIVION! (Warnock 1982, 22)

War involves rape literally, as well as symbolically, and militaries and militarism bolster patriarchal states in both war and what now passes for peace. Violence in war, violence in the street, and violence in the home are therefore inseparable.

Despite the nearly universal association of masculinity with war and violence, it need not be taken as evidence of any special *male* propensity for mayhem. Virtually all violence is *male* violence, to be sure, in the sense that men commit it. The cause of violence, however, is not the male sex. Rather, violence is necessary to the patriarchal system through which mostly white, privileged males hold power over others.

This violence is thus *patriarchal* violence. Violence in the service of domination is structural and is essential to patriarchy. War and the preparations for war are strategies of patriarchal domination as surely as rape is. The ever more militarized "war-at-a-second's-notice" system under which we all now live is *needed* to secure white-male-minority regimes against the claims of women and the Third World.

PATRIARCHAL DOMINANCE AND FEMINIST POWER

Under patriarchy, power is defined as dominance—that is, one person's ability to exert control over another person. Within the system, it is men's control over women. *Over* is the key word. The dominant model of power is based on the assumption that some people are better than others and, therefore, are more valuable and important. In the patriarchal system men are assumed to be more valuable than women just as in racist systems whites are considered to be more valuable than people of color. In all systems of dominance, the dominant group is presumed to be more valuable, more important, than subordinates. Not only are subordinate groups less valuable, but the dominant group assumes that subordinates' needs are less valid than its own. This assumption justifies the dominant group's imposition of control in order to meet its own needs at the expense of subordinates. In short, systems of dominance represent an ideological rationalization for oppression and the denial of human choice and development.

Assumptions of inequality grow from dualism—thinking that separates. In *Dreaming the Dark*, Starhawk has called this way of thinking the "consciousness of estrangement," that is, the idea that "the world is made up of separate, isolated, non-living parts, that have no inherent value. Inherent value, humanness, is reserved for certain classes, races, for the male sex; their power over others is thus legitimized" (5–6). Those in power assume that the prevailing order is right and good, not only for themselves, but also for their subordinates. Nevertheless, in a structure of inequality, many people's needs are not met while a few people's needs are met in superfluity and at the expense of the many. Differences in needs and the ability to meet those needs creates conflict. In *Toward a New Psychology of Women*, however, Jean Baker Miller points out that

> [w]ithin a framework of inequality the existence of conflict is denied and the means to engage openly in conflict are excluded. Further, inequality itself creates additional factors that skew any interaction and prevent open engagement around real differences. Instead, inequality generates hidden conflict around elements that the inequality itself has set in motion. In sum, both sides are diverted from *open conflict around real differences*, by which they grow, and are channeled into hidden conflict around falsifications. (1976, 13; emphasis added)

Conflict is not inherently negative. Conflicts, or differences, arise when individuals or groups have discordant needs. The conflict itself is neither bad nor good, and the method for resolving the conflict might be either destructive or constructive. In systems based on the dominance of one group over another, conflicts represent a potential threat. If inequality were exposed as the root cause of the conflict, then the dominant system's foundation would be threatened. And when conflict does surface, those in power do not seek to uncover the cause of the conflict and then resolve the problem; rather, the dominant group imposes its needs by *winning* the conflict. Conflict thus becomes competitive, and potentially violent. Both subordinate and dominant groups learn to avoid conflict based on genuine differences and, in fact, come to view it as inherently destructive.

When subordinate groups succeed in exposing basic inequities, they are named responsible for creating conflict. Miller explains that "if subordinates do not accept their place as inferior or secondary, they will initiate open conflict. That is, if women assume that their own needs have equal validity and proceed to explore and state them more openly, they will be seen as creating conflict" (1976, 17). Subordinate groups who uncover inequities that are fundamental to maintaining a system of dominance threaten the presumption of justice that props up the system.

Dominant groups do not want to remedy conflicts by eliminating inequities. Rather, they exert power in order to control subordinates and thereby "resolve" the conflict.

Dominants do not seek to resolve problems, their goal is to *win*. To maintain control. To stay on top of the hierarchy. Under dominance, patriarchy, war is inevitable.

Feminism offers an alternative to dominance and violence. Since feminism assumes that all people are equally valuable, there is no justification for dominance. Feminists believe that no one is more valuable than another, and no one's needs are more valid than another's.

The feminist model of power assumes that nothing stands in isolation, all is connected. Starhawk calls this view the "consciousness of immanence" in which "the world and everything in it is alive, dynamic, interdependent, interacting, and infused with moving energies: a living being, a weaving dance" (9). In the feminist model, information is shared, not hoarded by a few. Instead of a few members of the system having access to information and power over those who do not have access, information is available to anyone who wants or needs it. Moreover, conflict is not avoided. Open conflict and discussion offer an opportunity to address problems, an opportunity for growth through resolving differences. Conflict resolution can be time-consuming, and sometimes it is painful, but it does not rely on coercive violence or domination.

Feminist power is cooperative. It is power to effect a positive solution using individual choice. Those involved in conflict resolution have the ability to make decisions for themselves and to control their own lives. In the feminist model, choice and process are vital. The goal of the feminist model is to resolve conflicts; winning is an irrelevant concept.

APPLIED FEMINISM: CREATING CHOICES

Feminist process must be integrated into our lives on the individual, institutional, and global levels; it is not an addendum. To change the direction of the future, to reduce our reliance on military presence and solutions, feminism must replace patriarchy as a system of behavior. Such a system would resolve conflicts with respect to everyone's needs and would not rely on power over others, or domination, to succeed.

Feminists have begun to use our power to change institutions that rely on dominant power. We have recognized the role of domination and violence in many homes and have created the shelter movement in response. In most cities, shelters exist that offer refuge and support. In addition, in these shelters women are learning to make choices about their lives and to take control using feminist power. Along with individual support, feminists have educated the police and the courts and have lobbied for laws that make abusive behavior between partners a crime.

In the past women have not had access to information about their bodies, specifically their reproductive systems. Information remained in the hands of the patriarchal medical system, allowing that system to have power over women. By saying that each of us has the right to control our own bodies and to gain information about our bodies, the women's health movement has made basic changes in the medical system. Feminist health centers, the Boston Women's Health Book Collective, and self-help classes and centers have clearly stated that, in order for each of us to make choices about, or have control of, our bodies and our health, we must have information. Information should no longer remain only in the hands of the "professionals." Empowering women, lay and professional, to ask questions has shown many women that power can come from within each of us as we "take on" the medical establishment. By using newly found information to expose the number of unnecessary hysterectomies, D&C's (dilation and curettage), and cesarean sections, an accepted form of violence directed at women, the women's health movement has changed ways that health care is practiced and delivered.

Feminism stresses the importance of process, of recognizing the needs of and differences between each of us. Often the approach is difficult and we might compromise our commitment to process in order to achieve a particular end. In May 1985 at a National Abortion Rights Action League (NARAL) speak-out in Washington, D.C., women from fifty states publicly told their personal abortion stories and read letters from over forty thousand people. The media coverage was excellent, the organization outstanding. Certainly, the event was successful in its goal of focusing on women and their lives. (Those who are "anti-choice" focus always on the fetus but overlook the relevance of women's voices.) Yet, in order to attract public attention, "media stars" were developed and encouraged; those women or couples who told the more dramatic stories were highlighted to get media attention. Playing the media game forced these feminists to create a hierarchy, at times ignoring process because the end was so vital to each of us. We need constantly to ask ourselves whether we are compromising our process, which is essential to peaceful solutions, in order to achieve a short-term goal.

Within the peace movement we are seeing many attempts to use feminist strategies to bring about change. It has been a slow and difficult process. Many of us remember working against the Vietnam War, listening mostly to men speak out against the war, while we typed, baked brownies and cookies, and provided child care. The peace movement, in many ways, operated under the structure of patriarchal power. Decisions were made by a few, and an elite core of spokesmen became the leaders who spoke for everyone. By not attempting to change how we dealt with and defined power, we were not preventing future abuses of power and conflicts, either within the peace movement or globally.

For many feminists, the style of our peace work illustrates our broader definition of peace. At the Burlington Women's International League for Peace and Freedom, we are working with women who see peace being possible only when each of us recognizes, and is recognized for, her uniqueness and power within. To accomplish our goal of peace we have begun to incorporate the elements of feminist power into our process.

Contrary to many work settings in which information is hoarded by a few and some people read others' mail to gain access to information, we make an effort to share information. At most meetings we make time to share articles or particular knowledge we have acquired. Within this cooperative system, the group's power increases rather than one person having power over others. We make all decisions by consensus, acknowledging that each of us has the power to participate in decision making. Reaching consensus often takes time; however, we are usually assured that no one is feeling overpowered by the majority. Each of us learns to facilitate meetings by alternating this job, and the facilitator assures us that the feminist process is followed. She encourages everyone to participate, keeps the group focused, and makes sure that guests are clear about our discussions and activities. Many feminist groups are leaderless; many of us are able to speak with the media, and knowledge of the group and issues does not rest with just a few women. And we encourage other peace groups that we work with to use our principles. Through the process, a subtle change occurs; gradually, respect for individual opinions and perspectives grows.

Some feminists express concern that, as more feminists work in the peace movement, there will not be enough women working on women's issues. We can no longer separate feminist issues from peace issues. We cannot achieve peace without implementing feminist strategies into every aspect of society. Peace cannot be achieved when a few have power over many. Feminists must bring feminist strategies to all our work, on the local and global levels, making feminism, not patriarchy, the norm. Peace groups must support women's issues such as child care, equal pay, and Title IX of the 1972 Education amendment (prohibiting gender discrimination in higher education), since, without this equity, men will continue to hold power over women. Peace will not exist when some hold power and must fight to retain it.

REFERENCES

Brownmiller, Susan. 1975. *Against our will*. New York: Bantam Books.

Bulletin of the Atomic Scientist. 1975, November, 17.

Caputo, Phillip. 1977. *A rumor of war*. New York: Holt, Rinehart and Winston.

Daly, Mary. 1978. *Gyn/Ecology*. Boston: Beacon Press.

Elster, Ellen. 1981. Patriarchy—a working definition. In *Loaded questions*, edited by Wendy Chapkis. Amsterdam and Washington DC: Transnational.

Enloe, Cynthia. 1983. *Does khaki become you? The militarization of women's lives*. Boston: South End Press.

Gilligan, Carol. 1982. *In a different voice*. Cambridge, Mass.: Harvard University Press.

Miller, Jean Baker. 1976. *Toward a new psychology of women*. Boston: Beacon Press.

Reardon, Betty A. 1985. *Sexism and the war system*. New York: Teachers College Press.

Starhawk. 1982. *Dreaming the dark*. Boston: Beacon Press.

Warnock, Donna. 1982. Patriarchy is a killer. In *Reweaving the web of life*, edited by Pam McAllister. Philadelphia: New Society Publishers.

Women's Pentagon Action. 1981. Unity statement.

Zanotti, Barbara. 1979. Militarism and violence: A feminist perspective. Paper presented at the Riverside Church Disarmament Conference, New York.

♦ DORIS GRIESER MARQUIT AND ERWIN MARQUIT ♦

Gender Differentiation, Genetic Determinism, and the Struggle for Peace

Are women the peacemakers of the world? The question raises some philosophical and socioeconomic questions whose answers are important in developing strategies for the peace movement. It is a specific instance of the broader argument that women are inherently the possessors of certain "feminine" characteristics—and that these include valuable qualities and abilities as well as the traditionally attributed weaknesses and disabilities.

Attractive as the idea might be of seizing an old weapon often used against women and using it in their behalf, it is a strategy that lacks validity. As Janet Sayers has effectively argued,

> the celebration of a distinctive, supposedly biologically given femininity, and of women's "separate" sphere is reactionary. It may have been progressive and realistic in the past to hope to achieve sexual equality through gaining an equal valuation for women's role in the home with men's role in work outside the home. But the goal of achieving sexual

---

The authors had been invited to participate in the original Genes and Gender Conference held in New York in January 1986 but were unable to attend. While visiting professors at Humboldt University in Berlin, German Democratic Republic, in the fall of 1985, however, they were asked to discuss the role of women in the U.S. peace movement. The essay presented here is based on that discussion, which was published in slightly different form in *Wissenschaftliche Zeitschrift der Humboldt-Universität zu Berlin, 36*(4) (1987).

> equality on the basis of confining women and men to separate spheres of activity has now been rendered entirely anachronistic and futile by the historical development of the economy over the last several centuries. (1982, 190–91)

Our concern here is with the causes of certain observed differences between men's and women's grass roots participation in the international struggle for a peaceful world and the implications of our conclusions for future progress in many areas of human life. We need not pause over the fact that men have tended to hold the leading offices in large peace organizations, particularly those endorsed by governments, and have received over 90 percent of all Nobel peace prizes. Women have filled the ranks of participants and, certainly in the United States, have formed countless spontaneous peace groups.

In several international women's organizations, such as the Women's International Democratic Federation (and its U.S. affiliate, Women for Racial and Economic Equality) and the Women's International League for Peace and Freedom, working for peace has long had the highest priority. For a number of years public survey in the United States showed an increasing gap between the sensitivities of women and men to peace and certain other social issues. An understanding of the factors that have given rise to this gap is necessary if the participation of women in the organized peace movement is to be broadened and deepened. It is often implied, and sometimes argued directly, that women are genetically programmed toward peace, nonviolence, and the cherishing of life—that their biological role of bearing the young of the human species determines their social orientation. We shall argue that the biological factor plays at most a minimal role.

If a social, rather than biological, basis can be shown to give rise to this gender gap, then the possibility arises for deepening the sensitivities of both women and men to the very same factors to which women have been responding. We would in no way oppose the focus of women's groups, even of women-only groups, on peace—quite the contrary. The goal would be to close the gender gap by raising the commitment of men to the struggle for peace.

First we offer some evidence for the existence of the gender gap, then discuss the inadequacy of the biological factor as an explanation, and, finally, examine some of the social conditions in which a more reasonable explanation can be found.

The term *gender gap* has been widely used in the United States to designate observed differences between the opinions of men and women on political issues. Public opinion polls and election results began in the mid-seventies to reflect apparent differences in political attitudes of men and women on issues ranging from the environment, the death penalty, and social welfare programs to military spending, foreign intervention, and the threat of nuclear war. The concept serves as a useful starting point for an examination of some aspects of consciousness and activity that seem to be characteristic of women in ways that affect participation in a changing peace movement in the remaining years of this century.

Two factors, it has often been pointed out, have tended to keep political attitudes of men and women close together: the influence of social class and economic status on both, and the tendency of women to follow the lead of fathers and husbands in political matters. Indeed, a review of data from U.S. public opinion polls since the end of the Second World War has shown that, through the early 1970s, "male and female opinions were uniform on most issues" (Ridings 1982, 1). On two key issues, however, a significant gap could be seen as early as 1968: men approved of building the neutron bomb by a margin of 54 percent to 38 percent, while women disapproved by approximately the same margin. And more than two-thirds of women called the U.S. role in Vietnam a mistake, while only slightly more than half of the men concurred (Ridings 1982, 6). By the late 1970s polls were picking up a gap between men and women on a number of issues, including the use of force in foreign policy, the ability of the U.S. economy to achieve stable prosperity, and domestic issues grouped under the headings of "fairness" and "compassion."

The 1980 presidential election revealed a nine-percentage-point difference between the male vote for Reagan (56 percent) and the female vote for him (47 percent). As the participation of different groups in the electoral process was being studied, it should have become clear that the "women's vote" is not an undifferentiated, homogeneous entity. For instance, 20 percent more women from the South voted in 1980 than in 1952, while in other regions there was a slight decline. Which candidates and which issues benefited from this shift depended on how many of these Southern women were, for instance, black women influenced by the civil rights movement and how many were white women linked to right-wing fundamentalist churches. Political organizers have been quick to see that increasing voting by different combinations of groups could offer interesting possibilities. Bella Abzug has shown, for example, that Reagan's plurality in 1980 in the ten largest states was less than the number of all nonvoting white women in each of those states; less than the number of nonvoting black women and men in Michigan, New York, and California; and less than the number of nonvoting Hispanic women and men in California, New York, and Texas (1984).

An electoral gender gap clearly seems to have continued through the 1982 congressional elections, affecting a number of important races, and by September 30, 1984 the *New York Times* was saying: "Since 1980, when the voting patterns of the sexes emerged as a significant political factor, the partisan choices of men and women have diverged dramatically." Noting that "personality appeal" seemed to be playing a special role in the presidential race, the *New York Times* identified major party identification as the most significant gap: among women 46 percent called themselves Democrats, 25 percent Republicans, and 29 percent Independent (almost the same as four years earlier). Among men, on the other hand, 33 percent called themselves Democrats (a drop of 9 percent in four years), 30 percent Republicans, and 37 percent Independent.

On the issues in the 1984 elections (submerged beyond previous precedent by the triviality and superficiality of the campaign), women called the Reagan administration's handling of the economy "unfair" more often than men, and they often saw a risk of war in foreign policy that men considered "forceful." And we are reminded of the influence on poll results of the way questions are worded when we are told that, in a *New York Times* / CBS poll, 81 percent of men called Reagan "tough enough" to be president (1984).

The reelection of Reagan was a sharp disappointment to progressive forces in the United States in many ways. For a group like the Women's Vote Project, a coalition of fifty organizations formed in 1982 to raise the number of women registering and voting, the electoral results carried a particular message: the slogan "It's a man's world unless women vote" proved to need some rethinking, since the actual gender gap in the 1984 presidential race was only 4 percent, down from 9 percent four years earlier (*Women and the Vote*, 1984). According to *New York Times* exit polls, men voted 61 percent for Reagan, and women 57 percent (November 8, 1984). The Reagan election is clearly an example of the important role played by social factors such as class and race overriding to some extent the importance of gender differences.

The voting booth is only one place in which differences between men's and women's behavior can be manifested, however, and the struggle for peace has not taken place there alone. The peace movement in the United States has always seen itself, at least in part, as an extraparliamentary force, and the gender gap has been apparent not only during election campaigns but in response to calls for protests and actions around issues of threatened and actual use of military force by the United States. According to Emily DeNitto, "Women's thinking on specific issues, as well as their voting patterns, are part of this [gender] gap" (1984, M2). In October 1983, for example, a poll showed 46 percent of women approving of the U.S. invasion of Grenada, while an appalling 69 percent of men approved of it; only 39 percent of women approved of U.S. policy in Lebanon, while 53 percent of men (DeNitto 1984, M2). A national women's poll conducted by the Women's International league for Peace and Freedom during the first half of 1984 showed an overwhelming 78 percent supported the nuclear freeze, 65 percent favored decreased U.S. military involvement in Central America, and almost 60 percent wanted reduced military spending overall. In the same poll 70 percent of the women questioned supported increased spending for child care programs (Women's International League for Peace and Freedom 1984).

Women's peace activism in capitalist countries has a long and unbroken history. In the United States and several European countries the Women's International League for Peace and Freedom was formed to push for peace talks near the end of the First World War. The most dramatic actions in the peace movement in recent years have been taken by women and women's organizations. The Women's Peace Camp at Greenham Common, the site of a U.S. Air Force base in England, was founded in 1981 and served as a model for the Women's Peace Encampment in rural western New York state, which was established in 1983, and many local women's peace camps. These peace encampments have drawn younger women and radical feminists into the struggle against nuclear weapons; the Greenham Common camp broadened the original protest against NATO Cruise missiles to include racism, uranium mining, and waste of food surpluses. An attempt was made to turn the camp at the Seneca Army Depot in New York into a permanent feminist commune. During the 1960s U.S. women's peace groups were formed to oppose the war in Vietnam (such as Another Mother for Peace and Women Strike for Peace) and in England around the Aldermaston Easter marches. The Women's Pentagon Action consciously built on those models. Women's peace groups, slogans such as "Peace is a Women's Issue," and Mother's Day peace marches have been so conspicuous that the danger exists that the peace issue might seem to be exclusively a women's concern—especially in the United States and at the level of public concern if not policy-making.

Women's sensitivity to peace issues is often explained in biological terms. The fact that women bear, suckle, and raise children is said to make women more concerned about preserving life. The biological factor is introduced essentially in two forms: sociobiological and ethological.

In the sociobiological view, genetic factors not only assign women the physiological function of motherhood but also determine their gender roles in society (Wilson 1978). The sociobiological view attributes to evolutionary processes the genetic factors that lead to the so-called feminine tendency to protect human life—the product of motherhood. This "genetic imperative" generates a need on the part of women to hold a generally pacifist attitude that condemns and avoids all violence in order to protect their offspring. Men, on the other hand, are genetically conditioned to fight off other men who might impregnate a woman; therefore, men are aggressive and more prone to engage in war. According to sociobiology, men are genetically "programmed" to spread their genes by impregnating many women; if they should risk dying because of their aggression, their genetic imperative to protect the offspring that carry their genes thus is not endangered.

The other form of genetic determinism, ethology, asserts that human gender roles that are considered instinctive in female and male animals are the result of evolutionary processes (Lorenz 1966). Ethology is a forerunner of sociobiology, and it has essentially the same content as the sociobiological view. Both ethology and sociobiology are compatible with the concept of "natural law"—its source being either an expression of divine will or a component of an unchanging human nature.

The pseudoscientific nature of sociobiology is rooted in its mechanistic program to extend genetically determined behavior patterns of lower animals to human behavior. The qualitatively different roles played by human consciousness—social and individual—in conditioning and guiding human behavior are not taken into account by sociobiology. Genetic theories of human behavior are by their nature ahistorical, and they ignore the entire historical course of cultural development of human society. The impossibility of differentiating sociocultural factors from genetic factors has been adequately documented in recent critiques of the sociobiological theories of E. O. Wilson and the theories of racial differences in intelligence advanced by Arthur Jensen (1972, 1973) and Richard J. Herrnstein (1973).

If biological determinism is accepted as a valid approach, it then becomes socially necessary to shape the personalities of boys and girls to conform to what are now considered to be their natural gender roles. This viewpoint is dominant in all class-divided societies and persists in gradually weakening form in socialist societies as well, although the degree varies from country to country. Women are not only socially conditioned for the biological role of motherhood but also for assuming the principal responsibility for child care, shopping, preparation of food, clothing maintenance, house cleaning, and other domestic labor. Within the framework of this gender role, women, having the obligation of child care, are conditioned to be continuously alert and sensitive to the need for protecting the lives of their children. Their social conditioning must have some influence on attitudes to war and peace as well, however difficult it would be to assess this influence quantitatively. The feminist theorist Susan Griffin has noted the difference between "socially masculine and feminine values": "The roles society had given to men and women had produced different thinking and different ways of being in us. . . . Men, valuing power, produce nations, conflict and wars, and . . . women, valuing life, produce relationship, continuity and peace" (1982, 14–15).

The education of girls and boys in the United States strongly emphasizes the role of women as mothers and preferentially prepares men for service in the armed forced. The same educational system, however, seeks to glorify for both boys and girls every war in which the United States has been engaged during the last two centuries. Thus, cultural conditioning is not directed toward pacificism for women but, rather, against their direct use of force in favor of their support for the use of force by men. Passivity, lack of self-confidence, and looking to men for protection are the traits society has found to be feminine.

Under many circumstances, such education and social conditioning might be expected to keep the opinions of women and men reasonably similar, with neither challenging the dominant ideology. Indeed, prior to the late 1960s, public surveys did not show the significant gaps between men and women on foreign policy issues involving war and peace that they do today. Since that time a number of developments have occurred that have given rise to the gaps that are now being observed.

Women, who are disproportionately represented among those with the lowest incomes and living standards in U.S. society, have responded with strong objections to the rapid rise in federal spending on the military (doubled since 1980) and decreased spending for social services. The issue clearly has a class character and is of special concern to single women because of the likelihood that they will be poor. Female-headed families in the United States are four times as likely to be poor as male-headed or couple-headed families (Midgely 1985, 2).

The earnings gap between men and women in the United States has stubbornly resisted change through periods of depression and recovery, war and peace, and the two decades of the women's movement. Since 1930 employed women have averaged about 60 percent of men's earnings. Women and young children are increasingly the poorest people in the United States. In 1980 women living in poverty constituted 16.5 percent of the total female population; two of every three poor adults were women. Since the Second World War there has been a tremendous increase in the number of single-parent households headed by women, and since 1960 the number has doubled. Nearly 40 percent of these households and over 50 percent of those headed by African-American and Hispanic women live in poverty. Women's poverty is caused by continuing discrimination in education and employment and, for women of color, by racism; it accounts for the fact that the majority of recipients of federal programs for health, housing, and nutrition are women and the families they maintain (Midgely 1985, 2). Since 1981 these population groups have been the principal direct sufferers from the cuts in social programs in favor of increased military spending.

Under the Reagan administration, spending on arms was increased, while decreases occurred in the amounts spent on Aid to Families with Dependent Children (during Reagan's first term half a million families were completely cut from this program, 95 percent of them headed by single women), food stamps, school lunches, subsidies for child care, medical service, public housing, and other needed programs. The National Organization for Women (NOW) has pointed out that it is not just the poorest women who have been hit by the arms buildup but almost all employed women. Massive amounts of funds have been transferred from the civilian economic sector, where most women work, to military production. NOW estimates that every time the Pentagon's budget goes up a billion dollars (it asked for over twice that in 1984), 9,500 jobs for women disappear (DeNitto 1984, M2). Indeed, military spending creates fewer jobs than almost all other kinds of public and private spending.

The gender gap (in voting behavior and opinions) thus proves to be a more complex phenomenon than it seems at first. It is clear that a primary reason for differences between the lives and consciousness of women and men is the increased concentration of women (and their young children) at the lower levels of the socioeconomic scale. Such conditions have always been the rule rather than the exception for African Americans and some other minority groups, and within these groups women and children, while they are often the most severely affected, are not the only sufferers. Those who have exaggerated the importance and size of the gender gap, or who have seen it in isolation from other factors, have often ignored class differences among women. (They were also the most surprised that the gender gap did not play as significant a role as expected in the 1984 elections.) A study by the Women's International League for Peace and Freedom concludes:

> Although most working women (6 out of 10) earn less than $10,000, a significant number of women—most of them white—are economically comfortable. Recent Commerce Department statistics estimate that nearly

> one million women have an annual income of $30,000 or more and another six million women earn between $20,000 and $35,000. It is obvious that such different economic circumstances will produce quite different approaches to political issues and that it is poor women who have the most to gain from changed budget priorities and a progressive direction for the country. (Midgely 1985, 2)

It is not only the single woman heading a poor family who experiences the consequences of sharply distorted federal budget priorities. Even in the family with two parents present, the management of the household economy has traditionally been a woman's job; it is she who shops, cooks, and maintains clothing—makes the pay packet last. She is therefore closer to the economic problems created by the distortion of the economy by arms spending; she will feel most sharply a rise in prices and curtailment of services.

The preceding discussion demonstrates that social factors, especially the class status of women and the gender roles historically assigned to the female members of that class, predominate over biological factors in shaping the consciousness of women. The emphasis on biologically related arguments has negative effects in two directions. It deflects the women's movement from proper assessment of the class roots of women's oppression and weakens the ability of the peace movement to raise issues in a way that will be most effective for increasing the participation of women in peace activities.

Monopoly capitalism in the United States has always used racism to divide the working class, to create the illusion among white workers that they benefit from the lower wages paid to workers of color, so as to deflect them from a united struggle for higher wages and improved conditions of labor. Job segregation by gender and lower wages paid to women are used in the same divisive way. The burden that so obviously falls on women as a result of arms expenditures falls even more heavily on both male and female workers of color. Although various sections of the working class are differentially affected, the burden of the military budget is now depressing the living standards of the U.S. working class as a whole.

The women's movement is increasingly giving recognition to the common social roots of both racism and women's oppression, including theories of genetic determinism that form their ideological basis. The effective coupling of the antiracist and women's movements will necessarily bring together four principal struggles: against imperialism, war, racism, and women's oppression. This linkage was striking in Jesse Jackson's presidential campaigns. The recognized connection between the racist character of imperialism and the racism propagated by monopoly capitalism at home was responsible for the success of the campaign against Reagan's support of the South African apartheid regime. The struggle against this policy, one of the most significant developments in the United States in 1985, was shaped and guided primarily by leaders of the African-American community, who demonstrated their ability to draw large numbers of other Americans into the struggle. The facts that over 90 percent of African-American voters voted against Reagan in his last presidential race and that the Black Caucus in the U.S. Congress unanimously condemned the invasion of Grenada are powerful indications of the recognition that the principal victims of U.S. imperialism are not only the subjugated peoples of the developing countries but also the most exploited sectors of the U.S. working class (most estimates place 90 percent of African Americans in the working class).

The last two decades have been marked by steadily increasing numbers of unemployed in the United States, even at the peaks of the business cycle. Washington arbitrarily defines unemployment of 7 percent or more as full employment. The scientific and technological revolution continues to displace workers from jobs faster than it opens new jobs. Working-class women, as we have noted, and both women and men of racially and nationally oppressed groups bear a large share of the burden resulting from the arms budget. This burden is increasingly being extended to all workers, especially with the wave of plant closings as firms move their production out of the United States.

More and more, as men and women come to hold the same sorts of jobs, instead of leading effectively segregated working lives, their opinions on most issues will presumably differ less; the gender gap on peace issues, as on others, will tend to close. But the strength of peace forces will not necessarily increase unless the struggle against job discrimination based on gender is accompanied by a corresponding struggle against racism, since racism is still the principal weapon used by monopoly capitalism to divide the working class.

Finally, we should expect that, in the long run, greater social equality of men and women at the workplace will lead men to assume a greater responsibility in the home, as we see occurring already, especially among younger couples. This increased responsibility should raise the sensitivity of men to the effects of military expenditures on the cost and availability of child care, food, housing, and the like. A second long-term factor is the tendency toward more single-parent families. This tendency is due, in part, to the destructive effects of poverty on family life, although the increase in single-parent families is evident among higher-income families as well. It is also visible in socialist countries. As the economic independence of women increases, the necessity for women to remain in unsatisfactory relationships—one of the important factors that has served to hold the family together in bourgeois society—disappears. New family forms will undoubtedly emerge, and this development is perhaps the least adequately studied aspect of life today. Consequences of changing marital, family, and child rearing forms certainly deserve to be the subject of serious study.

The experiences of the peace movement over the last forty years have brought forth a variety of forms of struggle that were developed on the basis of the various concrete threats to world peace as they arose. Genetic determinism can deal only with human responses to threats to peace against a background of relatively unchanging biological factors. We have attempted to show that the strategy for the participation of women in the peace movement must be flexible in order to take into account short- and long-term changes in the material conditions of their lives.

REFERENCES

Abzug, Bella. 1984. *Gender gap: Bella Abzug's guide to political power for American women*. Boston: Houghton Mifflin.

De Nitto, Emily. 1984. Guardians of peace. *World Magazine* (March 10): M2.

Griffin, Susan. 1982. *Made from this earth*. New York: Harper and Row.

Herrnstein, Richard J. 1973. *I.Q. in the meritocracy*. Boston: Little, Brown.

Jensen, Arthur R. 1972. *Genetics and education*. New York: Harper and Row.

Lorenz, Konrad. 1966. *On aggression*. New York: Harcourt, Brace and World.

Midgely, Jane. 1985. *The women's budget*. Philadelphia: Women's International League for Peace and Freedom.

New York Times, 30 September 1984. "Diverging Politics of Sexes Seen in Poll," 14.

New York Times, 8 November 1984. "Exit Polls," 1, 11.

Ridings, Dorothy S. "The Gender Gap." Speech before the Cleveland City Club Forum, 24 September 1982.

Rosoff, Betty, and Ethel Tobach, eds. 1975–88. Genes and gender series. Vols. 1–5. New York: Gordian Press.

Sayers, Janet. 1982. *Biological politics: Feminist and Anti-feminist perspectives*. New York: Methuen.

Sociobiology: The debate evolves. 1981–82. *The Philosophical Forum* 13(2–3).

Wilson, E. O. 1978. *On human nature*. Cambridge, Mass.: Harvard University Press.

Women and the vote—1984: A fact sheet. 1984. Washington, D.C.: National Commission on Working Women.

Women's International League for Peace and Freedom. July 1984. Leaflet on national women's poll results.

♦ ETHEL TOBACH ♦

Discussion

Women and men are considered to be different because of their natures, social factors, or both. The word *nature* is used interchangeably with *biology, hormones, brains, genes*, and *heredity*; and the word *social* often stands for *culture, experience*, or *environment*. The general use of these terms stems from the acceptance of a dichotomy between some kind of "nature" and "nurture" underlying behavior. Because this dichotomy has been criticized by scientists, professionals, parents, and others, it has become customary to talk about the "interaction" of heredity and environment. Many of those who talk about interaction believe that in this way they can cancel out the possibility that doing research and theorizing about genes, or heredity, will be used to justify sexist and racist explanations of the differences among people. Unfortunately, application of the interaction explanation frequently leads to asking questions such as, How many of the differences in behavior, brain, gender, sex, and race are inherited, and how many are learned?[1] The other side of the question is, Can we do anything to change (improve) the differences if they are inherited, or if half or most of them are inherited?

Genetic determinist and interactionist approaches to differences between women and men have their expressions not only in racist and sexist policies and behavior but also in social movements for equality and justice. In the movement for gender equity, for example, there are genetic determinist explanations of women's leadership in peace movements—that women are peaceful by nature, for example, and therefore likely to participate in peace movements. Such rationales have been coupled with explanations of war as a result of a genetically determined "masculine nature." Genetic determinists make the morphological and physiological differences between women and men the basis for the fact that women and men can be socialized to behave differently. These perceived differences are then claimed to justify expectations that women and men will view social issues differently. Yet, having sperm or egg cells does not play a role in the decisions of a people or a nation to go to war. On the contrary, armed conflicts are derived historically from two societal processes: the struggle for equality and justice, and the struggle between peoples and nations for power and domination.

In attributing the reasons for war to biological processes, genetic determinists also fail to distinguish between these different struggles. In order to trace the cause of war, it is important to define what kind of war is at issue. For example, humanity has a rich history of resistance to slavery and injustice, either through peaceful resistance—deceptive accommodation or the nonviolent resistance advocated by Mahatma Gandhi and Martin Luther King, Jr.—or the application of physical force. These methods have been employed throughout human history by women and men. Historical conditions, not gender, have determined which were used and by whom. The confusion of physiological and societal processes weakens the struggle for equality, justice, and peace.

It is important to distinguish struggles for liberation from those that the peace movement seeks to abolish—that is, wars waged by those who seek to exploit others for power. Women and men have supported and waged these wars of conquest and have benefited from them. The gender of those who waged these wars depended on the type of societies in which they lived. Elizabeth I and Catherine the Great were military leaders in monarchic, imperial, feudal societies. Margaret Thatcher and Golda Meier were leaders in advanced, imperialist, capitalist societies.

Genetic determinists advance the argument that, in wars for power, men are the actual actors in the struggle for reasons of their "biology." The reasons given are: (1) physiological, morphological characteristics of strength and skill in the exercise of techniques that kill people; (2) quick responsiveness with anger and aggression; and (3) the inability to incubate and nurture the next generation. Moreover, according to these arguments, it is biologically more efficient to lose a certain number of men in wars than to lose a number of women whose physiology makes it possible for them to bring forth more people even after the biological father has been killed. Those who desire power in a society capitalize on the argument that men are inherently fit for waging war and develop techniques for encouraging the realization of this inherent quality. Whether one explains war as the "natural" outcome of evolutionary processes or the consequence of socialization, the equation becomes men = aggression = war.

This equations appears to present a conundrum that cannot be solved, a problem by which humanity must live and die. It has been suggested, however, that the solution lies in giving women power—since they would not use it to destroy lives. This solution is based on another expression of genetic determinism: since women incubate and nourish human progeny, they are incapable of destroying life.

In discussions of masculine aggression and feminine nurturance, exceptions to these stereotypes are excluded, except as "abnormalities." For example, in the United States today, most will accept the formulation that women who do not nurture children willingly are abnormal, atypical, or pathological. These so-called exceptions (and pathologies) are determined by society; other societies define abnormality differently. In parts of contemporary Brazil, for example, if a man kills another man because he believes his wife has been "unfaithful" (a behavior defined by the husband and not requiring objective proof), it is not considered pathological, abnormal, or atypical. In some Muslim societies, if a husband or father kills a wife or daughter because of unseemly behavior—for example, choosing her own husband in the case of a father and daughter—again it is not considered pathological, abnormal, or atypical. Alternatively, if an aristocrat or monarch gives her child over to a state-designated caregiver, wet nurse, or tutor and has little to do with the child except in formal situations, it is not considered a violation of the natural, nurturant, normal woman's psyche, nor is it considered abnormal. While some institutions in our society offer moral precepts of "love your brother" for men as well as women, men who do not wish to train for killing or for making weapons that kill are considered atypical.

The educational methods by which men are socialized to believe that they are natural war makers and women are socialized to want "their men" to be warriors while they keep the home fires burning and bear the children have been discussed in other sections of this volume. Also discussed are the attempts to justify military policies on the grounds of genetics and the physiological, behavioral expressions of those genetic processes. The rubric under which the papers in this section have been organized is their concern with the role of women in peace movements in human history. The story of *Lysistrata* has probably been played out in many periods and places of human activity. In this play by Aristophanes, Lysistrata proposes that the way to keep men from going to one of the many wars waged by ancient Greece is to refuse to have sexual intercourse with them. Similarly, the prevalent notion today is that the movements to prevent war and bring about an enduring peace are in the hands of the women of the northern and western hemispheres. Clearly, the women's movement for the cessation of war in Vietnam played a significant role in the eventual withdrawal of the United States from that country. It is equally evident, however, that this withdrawal came about through the confluence of the women's movement and the movement among young men of the United States to resist being drafted or participating in the war. In their resistance the women and youth were joined by fathers and other men who renounced U.S. policy in that genocidal action. It might be said that in the end it was the youth who were the critical factor in bringing the Vietnam War to an end.

It is also important to consider that, while women are active in antiwar movements, they are equally committed to struggle against tyranny wherever it exists, like their male counterparts who share this commitment. Women's liberationist zeal has been demonstrated throughout history and around the world—in the leadership of such women as Winnie Mandela in the struggle against South African apartheid; in the active planning and carrying out of military policy in Latin America in the pursuit of national liberation (by Haydee Santamaria in Cuba, Nora Astorga in Nicaragua, and the adolitas of Mexico, who fought with Emiliano Zapata and Pancho Villa); in the work of the Rani of Jhansi Regiment, the women's division of the Indian National Army, which fought against British imperial power; and in the revolutionary efforts of Clara Zetkin and Rosa Luxemburg in Germany.

The understanding that only historical, societal processes lead to the human activity known as war is derived from an appreciation of the relationship between human behavior and human physiology. All life is made possible by the integration of the biophysical, biochemical processes that include genes, hormones, enzymes, proteins, and other systems. We share these processes with all forms of life, animals and plants. As humans, however, we are characterized by a unique nervous system that evolved to produce human forms of thought, language, and consciousness. The species solved problems of environmental control for survival by transmitting knowledge through generations. As these uniquely human social activities developed historically, the control of humans by each other through the establishment of class structures affected the relationships of individuals who were different from each other by virtue of appearance or physiological sex. But, while individuals are engaged in complicated societal relationships, they continue to function physiologically, like all living organisms, and it is easy to confuse social and physiological explanations of their differences. The confusion is what causes genetic determinism to flourish, but blaming the differences between people on biology eliminates the need to consider the history of the consciousness of human beings.

The riddle of the relationship between human behavior and human physiology can be solved by understanding how their integration functions. The answer lies in putting the proper emphasis on those aspects of the integration required for resolving the problems of humanity: to cure diseases we need to rely on understanding physiological processes; to cure war we need to rely on thinking, language and consciousness.

NOTES

1. For my amplification of this discussion in other articles, see The meaning of cryptanthroparion, in *Genetics, environment and behavior*, edited by L. Ehrman, G. Omenn, and E. Caspari (New York: Academic Press, 1972); Femaleness, maleness and behavior disorders in nonhumans, in *Gender and disordered behavior*, edited by Edith S. Gomberg and Violet Franks (New York: Bruner/Mazel, 1977); Evolutionary aspects of the activity of the organism and its development, in *Individuals as producers of their development: A lifespan perspective*, edited by R. M. Lerner and N. A. Buch-Rossnable (New York: Academic Press, 1981); Evolutionary theories and the issue of nuclear war: Implications for mental health, *International Journal of Mental Health* 15, nos. 1–3 (1987): 56–64; Integrative levels in the comparative psychology of cognition, language and consciousness, in *Cognition, language and consciousness*, edited by G. Greenberg and E. Tobach (Hillsdale, N.J.: LEA, Inc., 1987); Biology and social sciences, *Nature, Society and Thought* 1(1988): 587–96.

Appendix

The document that follows, which has become known as the Seville Statement on Violence, was developed by a group of biologists and social scientists who were members or participants in a meeting of the International Society for Research on Aggression in May 1986. Since its formulation, it has been endorsed by many organizations: Psychologists for Social Responsibility; the American Psychological Association; the American Anthropological Association; the International Society for Research on Aggression; the Society for the Psychological Study of Social Issues; the American Association of Counseling and Development; and Movimiento por la Vida y la Paz (Argentina). It has served as a basis for course curricula, meetings and conferences, and peace research.

STATEMENT ON VIOLENCE

Believing that it is our responsibility to address from our particular disciplines the most dangerous and destructive activities of our species, violence and war; recognizing that science is a human cultural product which cannot be definitive or all encompassing; and gratefully acknowledging the support of the authorities of Seville and representatives of the Spanish UNESCO, we, the undersigned scholars from around the world and from relevant sciences, have met and arrived at the following Statement on Violence. In it, we challenge a number of alleged biological findings that have been used, even by some in our disciplines, to justify violence and war. Because the alleged findings have contributed to an atmosphere of pessimism in our time, we submit that the open, considered rejection of these misstatements can contribute significantly to the International Year of Peace.

Misuse of scientific theories and data to justify violence and war is not new but has been made since the advent of modern science. For example, the theory of evolution has been used to justify not only war, but also genocide, colonialism, and suppression of the weak.

We state our position in the form of five propositions. We are aware that there are many other issues about violence and war that could be fruitfully addressed from the standpoint of our disciplines, but we restrict ourselves here to what we consider a most important first step.

IT IS SCIENTIFICALLY INCORRECT to say that we have inherited a tendency to make war from our animal ancestors. Although fighting occurs widely throughout animal species, only a few cases of destructive intra-species fighting between organized groups have ever been reported among naturally living species, and none of these involve the use of tools designed to be weapons. Normal predatory feeding upon other species cannot be equated with intra-species violence. Warfare is a peculiarly human phenomenon and does not occur in other animals.

The fact that warfare has changed so radically over time indicates that it is a product of culture. Its biological connection is primarily through language which makes possible the coordination of groups, the transmission of technology, and the use of tools. War is biologically possible, but it is not inevitable, as evidenced by its variation in occurrence and nature over time and space. There are cultures which have not engaged in war for centuries, and there are cultures which have engaged in war frequently at some times and not at others.

IT IS SCIENTIFICALLY INCORRECT to say that war or any other violent behaviour is genetically programmed into our human nature. While genes are involved at all levels of nervous system function, they provide a developmental potential that can be actualized only in conjunction with the ecological and social environment. While individuals vary in their predispositions to be affected by their experience, it is the interaction between their genetic endowment and conditions of nurturance that determines their personalities. Except for rare pathologies, the genes do not produce individuals necessarily predisposed to violence. Neither do they determine the opposite. While genes are coinvolved in establishing our behavioural capacities, they do not by themselves specify the outcome.

IT IS SCIENTIFICALLY INCORRECT to say that in the course of human evolution there has been a selection for aggressive behaviour more than for other kinds of behaviour. In all well-studied species, status within the group is achieved by the ability to cooperate and to fulfil social functions relevant to the structure of that group. "Dominance" involves social bondings and affiliations; it is not simply a matter of the possession and use of superior physical power, although it does involve aggressive behaviours. Where genetic selection for aggressive behaviour has been artificially instituted in animals, it has rapidly succeeded in producing hyper-aggressive individuals; this indicates that aggression was not maximally selected under natural conditions. When such experimentally created hyper-aggressive animals are present in a social group, they either disrupt its social structure or are driven out. Violence is neither in our evolutionary legacy nor in our genes.

IT IS SCIENTIFICALLY INCORRECT to say that humans have a "violent brain." While we do have the neural apparatus to act violently, it is not automatically activated by internal or external stimuli. Like higher primates and unlike other animals, our higher neural processes filter such stimuli before they can be acted upon. How we act is shaped by how we have been conditioned and socialized. There is nothing in our neurophysiology that compels us to react violently.

IT IS SCIENTIFICALLY INCORRECT to say that war is caused by "instinct" or any single motivation. The emergence of modern warfare has been a journey from the primacy of emotional and motivational factors, sometimes called 'instincts,' to the primacy of cognitive factors. Modern war involves institutional use of personal characteristics such as obedience, suggestibility, and idealism, social skills such as language, and rational considerations such as cost-calculation, planning, and information processing. The technology of modern war has exaggerated traits associated with violence both in the training of actual combatants and in the preparation of support for war in the general population. As a result of this exaggeration, such traits are often mistaken to be the causes rather than the consequences of the process.

We conclude that biology does not condemn humanity to war, and that humanity can be freed from the bondage of biological pessimism and empowered with confidence to undertake the transformative tasks needed in this International Year of Peace and in the years to come. Although these tasks are mainly institutional and collective, they also rest upon the consciousness of individual participants for whom pessimism and optimism are crucial factors. Just as "wars begin in the minds of men," peace also begins in our minds. The same species who invented war is capable of inventing peace. The responsibility lies with each of us.

Seville, May 16, 1986

DAVID ADAMS, Psychology, Wesleyan University, Middletown, CT, USA
S. A. BARNETT, Ethology, The Australian National University, Canberra, Australia
N. P. BECHTEREVA, Neurophysiology, Institute for Experimental Medicine of Academy of Medical Sciences of USSR, Leningrad, USSR
BONNIE FRANK CARTER, Psychology, Albert Einstein Medical Center, Philadelphia, PA, USA
JOSÉ M. RODRÍGUEZ DELGADO, Neurophysiology, Centro de Estudios Neurobiológicos, Madrid, Spain
JOSÉ LUIS DÍAZ, Ethology, Instituo Mexicano de Psiquiatría, Mexico D.F., Mexico
ANDRZEJ ELIASZ, Individual Differences Psychology, Polish Academy of Sciences, Warsaw, Poland
SANTIAGO GENOVÉS, Biological Anthropology, Instituto de Estudios Antropolóqicos, Mexico D.F., Mexico
BENSON E. GINSBURG, Behavior Genetics, University of Connecticut, Storrs, CT, USA
JO GROEBEL, Social Psychology, Erziehungswissenschaftliche Hochschule, Landau, Federal Republic of Germany
SAMIR-KUMAR GHOSH, Sociology, Indian Institute of Human Sciences, Calcutta, India
ROBERT HINDE, Animal Behaviour, Cambridge University, UK
RICHARD E. LEAKEY, Physical Anthropology, National Museums of Kenya, Nairobi, Kenya
TAHA H. MALASI, Psychiatry, Kuwait University, Kuwait
J. MARTIN RAMÍREZ, Psychobiology, Universidad de Sevilla, Spain
FEDERICO MAYOR ZARAGOZA, Biochemistry, Universidad Autónoma, Madrid, Spain
DIANA L. MENDOZA, Ethology, Universidad de Sevilla, Spain
ASHIS NANDY, Political Psychology, Center for the Study of Developing Societies, Delhi, India
JOHN PAUL SCOTT, Animal Behavior, Bowling Green State University, Bowling Green, OH, USA
RIITTA WAHLSTRÖM, Psychology, University of Jyväskylä, Finland

Notes on the Contributors

LINNDA R. CAPOREAL is Associate Professor of Psychology in the Department of Science and Technology at Rensselaer Polytechnic Institute and a member of the Genes and Gender Collective. She is a contributor to *Science*, the *Journal of Social Issues*, the *Journal of Personality and Social Psychology*, and *Behavioral and Brain Sciences*. Her current research interests include the social psychology of human-computer interactions, the evolution of human sociality and cognition, and cognitive heuristics and biases in decision making. She is currently working on a book on the evolution of social cognition.

GERALDINE J. CASEY is a doctoral candidate in anthropology at the Graduate Center of the City University of New York (CUNY). She is currently working to complete her dissertation project, an ethnographic study of popular culture and class consciousness among women clerical workers in Puerto Rico. A member of the Genes and Gender Collective, Geraldine had the opportunity to work with Eleanor Leacock as a teacher in the City College Anthropology Department. While at City College, she organized one of the first public forums where Leacock presented her Samoan research. On the basis of this working relationship and her respect for Leacock's Marxist-feminist scholarship, Geraldine accepted the task proposed by this volume's editors that she present Leacock's anthropological perspective on warfare and her critique of Marvin Harris.

ELLEN DORSCH is an independent health planning and policy consultant working primarily in Vermont. Her most recent work includes a report on the status of prenatal care in Vermont, program development in family planning and maternal and child health, and a chapter on heart disease recently published in *Ourselves Growing Older*, a book for mid-life and older women. She is a health activist, working particularly toward a state plan in Vermont for abortion rights, and in community health. She is also a member of the Burlington, Vermont, chapter of the Women's International League for Peace and Freedom.

CATHERINE M. FLAMENBAUM is a Ph.D. candidate in social psychology at the State University of New York at Stony Brook and a feminist activist. She is currently working as an educator, program designer, and caregiver in the field of early childhood education. Her research interests include the psychology of antisovietism, the relationship between feminism and peace activism, and child maltreatment: social roots and societal responses. She is a member of the Genes and Gender Collective.

DAISY GARTH grew up in Bluefields on the east coast of Nicaragua and moved to Managua in 1983. There she studied agricultural economies at the National Autonomous University and developed an interest in women's issues.

SUSAN G. GORDON was Associate Professor of Clinical Pediatrics at Columbia University and an attending pediatrician at Columbia Presbyterian's Babies Hospital; St. Luke's Roosevelt Hospital; and Harlem, Metropolitan, and Flower–Fifth Avenue hospitals. From 1952 to 1956 she was codirector, with her husband Edmund W. Gordon, of the Harriet Tubman Child Health and Guidance Clinic in New York City. Susan is a member of the Genes and Gender Collective.

ANNE E. HUNTER is a feminist community activist and a social psychologist in the Department of Psychology at Queens College, the City University of New York. As part of her applied work in psychology, she has developed an alternative high school program in Williamsburg, Brooklyn, for youth in poverty: You Can Community School; and founded Wingspan, a wilderness, survival camp in northeastern Tennessee for the physical, psychological, and spiritual development of youth in poverty. Her current research interests are in the areas of the mental health of women in a sexist society, the psychological empowerment of socially oppressed groups, and the interrelationships among social conditions, the psyche, and immunity. She is a member of the Genes and Gender Collective.

MAMIE JACKSON is an activist who has been involved in community and peace activities for most of her life. For thirty years, she worked as a tenant organizer for the Bronx Council on Rents and Housing and subsequently, as an advocate for senior citizens with the New York City Housing Authority's Department of Aging. She currently lives in Seattle, writing on the issues that concern her most.

SALLY L. KITCH is Associate Professor of Women's Studies and Director of the Center of Women's Studies at Wichita State University. She is the author of *Chaste Liberation: Celibacy and Female Cultural Status*, winner of the National Women's Studies Association Book Award. Her main research interest is the relationship between gender and culture.

JOY LIVINGSTON is Assistant Research Professor in the Department of Psychology at the University of Vermont. She is Project Director at the Center for Community Change through Housing and Support, a professor of women's studies, a feminist activist interested in issues of power, peace, and freedom from violence, and a member of the Burlington, Vermont, chapter of the Women's International League for Peace and Freedom.

DORIS GRIESER MARQUIT is on the adjunct faculty in the Women's Studies Program and the Program in Creative and Professional Writing of the English Department at the University of Minnesota. She is managing editor of the Marxist Educational Press.

ERWIN MARQUIT is a professor in the physics department at the University of Minnesota, where he has taught since 1966. He is the editor of *Nature, Society, and Thought: A Journal of Dialectical and Historical Materialism*.

SUSAN OYAMA is Professor of Psychology at John Jay College, the City University of New York. She is interested in the nature-nurture opposition, and, more generally, in the use (and abuse) of biological ideas in the social sciences, in biology itself, and in society at large. Her writings include articles on the concept of development and the relationship between development and evolution; a book on the concept of genetic information called *The Ontogeny of Information: Developmental Systems and Evolution*; and a coauthored and coedited book called *Aggression: The Myth of the Beast Within* (under the pen name John Klama).

MARY BROWN PARLEE is Professor of Psychology at the Graduate School and University Center of the City University of New York (CUNY). She was Director of the Center for Study of Women in Society at CUNY from 1979 to 1984 and is a past president of the Division of Psychology of Women of the American Psychological Association. Her current research interests are the psychology of menstruation, premenstrual syndrome, and eating disorders.

JOANNA RANKIN is a professor of physics and astronomy at the University of Vermont. As part of her study of pulsars she has initiated international cooperative research projects in India, the Soviet Union, and Poland. She is a member of the Burlington, Vermont, chapter of the Women's International League for Peace and Freedom. She is a feminist activist involved in issues of peace and reproductive choice.

BETTY ROSOFF, an endocrinologist, is now Professor Emerita in the Biology Department at Stern College of Yeshiva University. She is a member of the International Committee against Racism and serves on the editorial board of the journal, *The Prostate*. She applies her expertise in endocrinology to the investigation of gender and biological determinism. Betty is a founding member of the Genes and Gender Collective.

GEORGINE SANDERS [Vroman] was born in Indonesia, studied medicine in Utrecht, the Netherlands, and lived there during World War II under the German occupation. She has lived in the United States since 1947 and is now a medical anthropologist working in cognitive rehabilitation of the elderly with memory problems and with brain injury survivors. She is a consultant at the Geriatric Clinic of Bellevue Hospital Center, New York, and an associate of Cognitive Rehabilitation Services, Sunnyside, New York. In recent years she has been publishing poetry under the name of Georgine Sanders in the Netherlands and the United States. Georgine is a founding member of the Genes and Gender Collective.

SUZANNE R. SUNDAY is a Research Associate at Cornell University Medical College in White Plains, New York. Her current research focuses on meal eating, food preferences, and attitudes toward food in people with eating disorders. Previously, she taught psychology and conducted research in feeding behavior in a variety of rodent species.

ETHEL TOBACH is a comparative psychologist at the American Museum of Natural History and Adjunct Professor in Biology and Psychology at the City University of New York. Her research deals with evolution and the development of social-emotional behavior. She has written extensively on the role of science in societal processes leading to racism and sexism. Ethel is a founding member of the Genes and Gender Collective.

REGINA E. WILLIAMS is a poet and writer. Her work has appeared in such publications as *New City Voices: An Anthology; Long Journey*

Home: An Anthology; Confirmation: An Anthology of African American Women; Drum; and *Black American Literature Forum*. Her nonfiction publications include *My Village, My Country, My World; Leadership*, two minicourses for black young adults; and *Our Work and God's World*, a racial/ethnic identity text for young adults. She is a founding member of Metamorphosis Writers Collective and the Ain't I A Woman Writing Collective and a member of the Genes and Gender Collective. In addition, she is pursuing an M.A. in Applied Anthropology at City College, CUNY where she is doing research in the area of women and children in society.

JUDITH WISHNIA is Associate Professor of Women's Studies and History at the State University of New York at Stony Brook. She has published a book, *Development of Class Consciousness and Unionization among French Civil Service Workers* (Baton Rouge, 1990) and articles on women civil service workers and women in World War I. Her current research interests are French labor history and women's political history.